Vol. VIII, No. 2.   April 1, 1937

University of Arizona
Bulletin

ARIZONA BUREAU OF MINES

ARIZONA GOLD PLACERS
AND PLACERING

(Fourth Edition, Reprinted 1937)

ARIZONA BUREAU OF MINES, MINERAL TECHNOLOGY SERIES
NO. 38. BULLETIN NO. 142.

PUBLISHED BY
University of Arizona
TUCSON, ARIZONA

PREFACE TO REPRINT OF FOURTH EDITION

It is unfortunate that it is necessary to reprint Bulletin No. 135 without revising it and bringing it up to date, but to have done so would have delayed its appearance about a year.

The completion of the report on the geology and ore deposits of the Tombstone district has taken all the time of the only field geologist employed by the Bureau, for many months, and it was believed to be undesirable to interrupt that work.

There has been a greater demand for this bulletin than for any other issued by the Bureau, and this demand is still very strong. The only way in which this demand can be met is to reprint the bulletin issued four years ago. It is doubtful if many months of field work would reveal changes of material significance since no important new placer field has been discovered in Arizona since Bulletin No 135 was prepared.

G. M. BUTLER

June 3, 1937

PREFACE TO FOURTH EDITION

The first publication of the Arizona Bureau of Mines on Arizona Gold Placers was written by Mr. M. A. Allen, when Geologist with the Bureau, and appeared in 1922 as Bulletin 118. It was mainly a compilation of data already in print, but scattered and difficult to find. The stock of this bulletin was exhausted in four years. Mr. Eldred D. Wilson, Geologist with the Bureau, was then commissioned to re-write the bulletin, adding what new data could be obtained. Assisted by Mr. W. R. Hoffman in the field and further aided by the advice and suggestions of Dr. Carl Lausen, then Geologist with the Bureau, Mr. Wilson completed his work in the summer of 1927 and a large edition was published at once at Bulletin 124. Conditions with which everyone is familiar developed within two or three years and so much interest was shown in gold that the demand for this bulletin was extremely heavy and the supply was exhausted before June of last year. A new and greatly enlarged bulletin, No. 132, was prepared at that time, and, although five thousand copies were printed, they have all been distributed.

The present bulletin does not differ radically from No. 132, but many parts of that bulletin have been re-written and an attempt has been made to bring the information obtained therein up to date. Mr. Eldred D. Wilson made the field investigations required to secure this new data, and at least a dozen additional districts are described in the present bulletin.

Parts II to VII, inclusive, have been presented by Mr. G. R. Fansett to thousands of people who have attended short courses for gold prospectors, which he has conducted in centers of population all over the state during the past year or two, and experience has shown that the information conveyed is very useful, especially to inexperienced persons.

With the exception of very recent discoveries, in spite of diligent efforts to gather all the information available, the descriptions of Arizona placer fields are incomplete and otherwise unsatisfactory. It could, however, hardly be otherwise. The pioneer prospectors and miners were too busy overcoming obstacles, struggling against hardships and celebrating occasional periods of good fortune to write about their experiences, even if able to do so. Few authentic records of most of the earlier camps exist. Available statistics are often far from reliable, and good judgment is required to separate the true from the false.

Anyone who secures a copy of this bulletin with the idea of obtaining therefrom such data as will enable him to engage

3

profitably in placer mining in Arizona should remember that gold placers are usually the first deposits found and exhausted in every region. Prospecting for placer gold is not expensive, and a deposit once found can be worked with little capital unless dredging is necessary. Even hydraulic operations (which are not described in this bulletin because it is doubtful if any deposits that can be worked satisfactorily in this way exist where the requisite water is available) do not ordinarily require the expenditure of any considerable sum for equipment unless the water must be piped or flumed long distances. Because placer gold can be easily and cheaply recovered where water is available, it is not likely that unworked ground of fairly good grade remains, at least along streams which flow for several months a year. People attempting to do placering in such districts must, therefore, ordinarily be satisfied to work ground where the difficulties encountered, such as the prevalence of huge boulders, were too great or the grade of the gravel was too low to attract the old-timers. Hundreds of people are, however, trying to earn wages on such ground now.

Although there is undoubtedly much placer gold in the so-called "desert" regions of southern Arizona, the lack of water, both for placering operations and for use in camp, is a serious drawback there, as are also the cemented conditions of the gravel in several areas. Many types of dry-washers have been tried in these regions, usually with very indifferent success for reasons outlined in this bulletin, and the high summer temperatures that prevail there should deter anyone from prospecting in these areas during the summer months unless he is accustomed to the conditions he will encounter and knows how to meet them.

Recent field investigations made by Mr Eldred D. Wilson reveal the fact that the average daily recovery of each experienced placer miner in the State is probably less than a dollar a day, while inexperienced persons are averaging less than 25c a day.

Of course these statements mean that a few are doing fairly well, a larger number are earning expenses, and the majority are not recovering enough gold to buy food. Rumors that good wages can be made in this way, therefore, should be heavily discounted. A person not in robust health or one who has not sufficient funds to finance his entire trip runs a splendid chance of starving to death if he tackles placer mining in Arizona. If, however, a man in good health is out of work, has enough money to pay camp expenses for some time, and is willing to work hard, a prospecting trip will doubtless prove preferable to lying around and doing nothing, but it should be taken with the full realization that it is highly probable that little gold will be found. Of course, some rich, virgin ground may be found, but the chance of making such a discovery is small. It is this chance, however, that has actuated all prospectors and led to the discovery of most mineral deposits.

G. M. Butler

August 15, 1933

4

CONTENTS

PART II

SMALL SCALE GOLD PLACERING

PART VI

PART VII

LIST OF ILLUSTRATIONS

PART I

ARIZONA GOLD PLACERS

(Fourth Edition, Revised)

By ELDRED D. WILSON

Geologist, Arizona Bureau of Mines

GENERAL ORIGIN AND FEATURES OF GOLD PLACERS

Gold placers, or deposits such as gravel and sand which comtain notable concentrations of gold, all result from the slow milling and concentration processes incident to the natural erosion of pre-existing gold-bearing rocks. The origin of many gold placers is traceable directly to auriferous veins, lodes, or replacement deposits which, in most instances, were not of high grade.

According to Emmons,[1] placers are not apt to form from gold-bearing outcrops that contain abundant manganese, iron sulphides, and chlorides, unless precipitating agents such as calcite, siderite, rhodochrosite, pyrrhotite, chalcocite, nepheline, olivine, or leucite are abundant, or unless erosion is very rapid. In other words, the gold may be dissolved and carried below by means of natural chlorination processes that the established when solutions containing chlorides, together with sulphuric acid from the oxidation of iron sulphides, act upon manganese dioxide; but this process is neutralized if precipitating agents are present, and may be ineffective if erosion is very rapid.

According to Lindgren,[2] the best conditions for the concentration of gold into placers are found where deep secular decay of the rocks has been followed by slight uplift. As the rocks of a region break up under weathering, rainfall washes away most of the resultant detritus, grinds it by striking and rubbing it together and by dragging it along the stream bed, and liberates most of the included gold. Because gold is six or more times heavier than ordinary rock, the liberated particles of gold will concentrate along the bottom and come to rest where the stream

---

[1] Emmons, W. H., The enrichment of ore deposits: U S Geol Surv. Bull 625, pp. 305-324. 1917.

[2] Lindengren, Waldemar, Mineral Deposits, 3d ed, pp. 245-268, 1928

gradient lessens. The coarser particles will settle down first, and the fine and flaky gold will be carried farther along. The best placer concentrations probably occur in rivers of moderate (about thirty feet per mile) gradient, under nicely balanced conditions of erosion and deposition. Except where gravel bars may form in certain slower reaches, particularly within the arcs of curves, very little concentration will take place in the gorges. Such bars, through further deepening of the channel, may be left as elevated benches.

Most of the gold in a placer usually rests on or near the bedrock. In some instances, the coarser gold is scattered through the lower four to twenty feet, or the gravel may be richest a few feet above bedrock, but never is the richness equally distributed vertically. Among the best types of bedrock are compact clays, somewhat clayey, decomposed rock, and slates or schists whose partings from natural riffles. Smooth, hard material does not catch or retain the gold effectively. Gold works down for some distance into the most minute crevices of hard rock, for one to five feet into the pores of soft rock, and for many feet along the solution cavities of limestones.

According to Lindgren,[3] crystallized gold, which is sometimes found in placers, indicates close proximity to the primary deposit. He states that there is probably no authenticated case of crystallized gold occurring in gravels which have been transported far, and that it is difficult to believe the assumption that such crystals are formed by secondary processes in the gravels. The high insolubility of gold in most surface waters is demonstrated by the fact that flake or flour gold, which commonly is in 2,000 particles per one cent's worth, may be carried by rivers of moderate gradient for hundreds of miles.

The fineness, or parts of unalloyed gold per thousand, of placer gold is usually greater than that of the vein gold of the same district. This increase in purity, which is proportional to the distance that the placer material has been transported and to the decreasing size of the grains, has been shown to be due to the solution and abstraction of silver by surface waters.

GENERAL DISTRIBUTION OF ARIZONA GOLD PLACERS

Due to the presence of gold-bearing rocks in most of the mountain ranges of the Southwest, gold placers which have been of economic importance occur in every county of Arizona except Apache, Coconino, and Navajo. As indicated on the accompanying map (Figure 1), the placer districts of Arizona that have been notably worked are in the southwestern mountainous and desert half of the state. Many additional placers, not of economic importance, occur in the gulches that issue from the numerous mineralized areas throughout this region.

---

[3] Lindgren, Waldemar, work cited

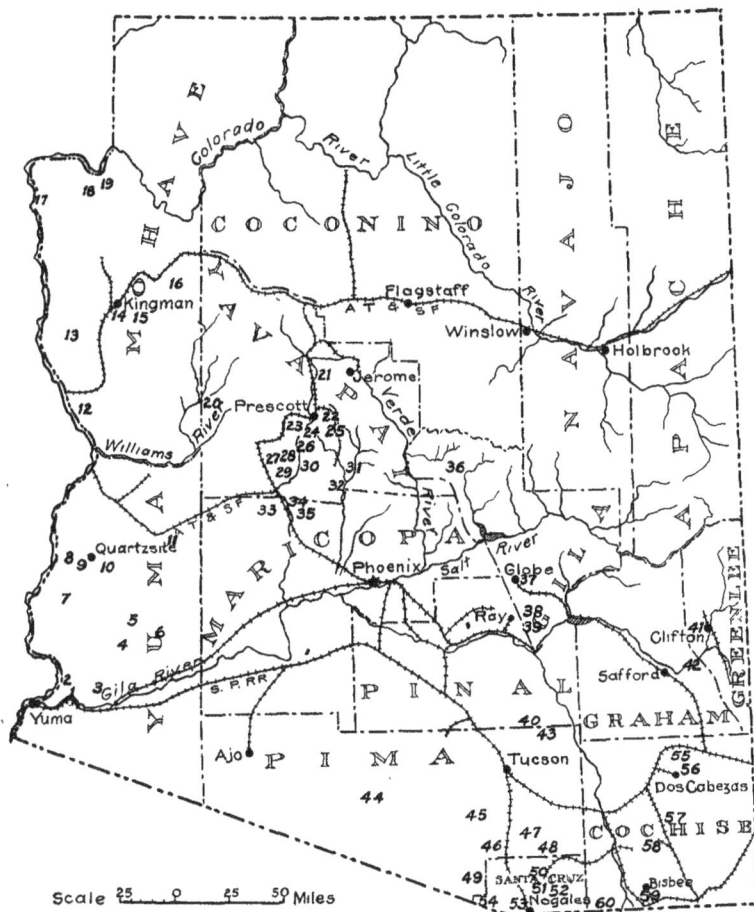

Figure 1. Index map showing location of Arizona gold placer districts.

1. Gila City (Dome).
2. Laguna.
3. Muggins.
4. Castle Dome.
5. Kofa or S. H.
6. Tank Mountains.
7. Trigo.
8. La Paz.
9. La Cholla, Oro Fino, Middle Camp.
10. Plomosa.
11. Harquahala.
12. Chemehuevis.
13. Silver Creek.
14. Lewis.
15. Lookout.
16. Wright Creek.
17. Willow Beach.
18. Gold Basin.
19. King Tut.
20. Eureka.

21. Granite Creek.
22. Lynx Creek.
23. Copper Basin.
24. Groom Creek.
25. Big Bug.
26. Hassayampa (Yavapai County).
27. Model.
28. Placerita.
29. Weaver, Rich Hill.
30. Minnehaha.
31. Black Canyon.
32. Humbug.
33. Vulture.
34. Hassayampa (Maricopa County).
35. San Domingo.
36. Payson.
37. Globe-Miami.
38. Dripping Spring.
39. Barbarossa.

40. Cañada del Oro.
41. Clifton-Morenci.
42. Gila River.
43. Alder Canyon.
44. Quijotoa.
45. Papago.
46. Armargosa.
47. Old Baldy.
48. Greaterville.
49. Las Guijas or Arivaca.
50. Tyndall.
51. Harshaw.
52. Patagonia or Mowry, Palmetto.
53. Nogales.
54. Oro Blanco.
55. Teviston.
56. Dos Cabezas.
57. Pearce.
58. Gleeson.
59. Gold Gulch (Bisbee).
60. Huachuca.

RELATION TO PHYSIOGRAHPY

Relation to pediments: A pediment, as defined by Bryan[4] is a more or less hilly plain, carved on solid rock and largely without alluvial cover, at the base of a desert mountain. The mountain slopes of a semi-arid region tend to have a steep profile that is uninterrupted by notably flat or gently sloping areas except where the bedding dips at low angles. Most of the dissected basins or flats that interrupt steep mountain slopes in this region prove, upon analysis, to be related to elevated pediments.

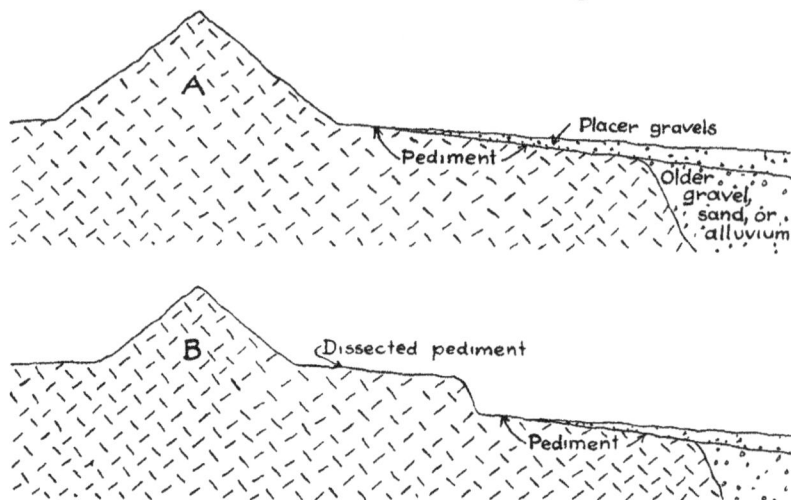

Figure 2. Ideal cross-sections of mountain pediments (B) represents the effect of renewed uplift and long erosion upon (A). the pediment of (A) has been more or less dissected, and a newer one has been formed at the base of the mountain.

The gold placers of Arizona, with the possible exception of a few that occur within mountain valleys or gulches, are related to pediments. The gold-bearing gravels occur not only in gulches and old channels which traverse or issue from pediments, but also, in many cases, as mantle upon the pediment itself. This relation may be explained as follows: As stated on page 9 the best conditions for the concentration of gold into placers are where deep decay of the rocks has been followed by slight uplift. When removed by erosion, decayed rocks tend to liberate their heavy minerals within sufficiently short space to promote concentrations. Undecayed rocks, on the other hand, are broken up by mechanical erosion which does not tend to release the heavy minerals with sufficient uniformity to produce placer deposits. In arid regions, mechanical erosion generally keeps ahead of notable rock decay on steep slopes but falls behind such decay on the gentle slopes of pediments.

[4] Bryan, Kirk, U S Geol. Survey Water-Supply Paper 499, p. 93. 1925.

Relation to streams: Most of the streams that have formed gold placers in Arizona were small, intermittent, and subject to torrential floods. Hence, placers of economic importance have been found to extend only for a relatively short distance downstream from mineralized pediment areas. Because of the intermittent character of the streams, many of the placers contain part of their gold more or less erratically distributed through a considerable thickness of gravel. In general, however, the richest material occurs at or near bedrock, especially where the bedrock surface forms natural riffles or contains irregularities such as potholes.

Particularly along some of the larger streams, notable placers occur as elevated bars which were deposited within the inner arcs of curves.

RELATION TO REGIONAL GEOLOGY AND TO TYPES OF VEIN DEPOSITS

The major gold placers of Arizona are associated with areas of crystalline rocks, such as schist, granite, and gneiss, where the veins tend to be of the deeper-seated mesothermal and hypothermal types.

Although many areas of volcanic rock occur throughout the State and in places contain abundant gold-bearing veins of the shallow-seated or epithermal type, none of them have given rise to gold placers of economic importance.

YEARLY RAINY SEASONS OF ARIZONA

The advent of rain is of great importance to the placer miner in Arizona   It exposes nuggets and provides temporary water for wet methods of concentration, put it hinders the dry-washer, whose dirt must be dry. Usually in Arizona, as in much of the Southwest, the least rain falls in May and June and the most during July, August, and the winter. Often this rain comes with a local violence that fills dry arroyos with torrents.

HISTORY OF ARIZONA GOLD PLACER MINING

The original discovery of placer gold in Arizona probably was made by Indians long before the advent of white men. As early as 1774, according to Elliot's History of Arizona (1884), certain placers of the Quijotoa district, about seventy miles west of Tucson, were being worked extensively by Padre Lopez, a Castilian priest   In 1858, according to Hamilton,[5] placers were dis-

---

[5] Hamilton, Patrick, Resources of Arizona. 1883.

covered on the Gila River, about twenty miles east of where it joins the Colorado, by Col. Jacob Snively. About 1862, the La Paz placers, near the Colorado River about 65 miles north of Yuma, were discovered by Capt. Pauline Weaver. The greatly increased prospecting that followed these discoveries soon resulted in the finding of the Dome Rock, Plomosa, San Domingo, and Yavapai County gold gravels. The Greaterville placers became known in 1874, and by 1900 many additional, but less important, discoveries were made in various parts of the State.

Since the most important placer fields of Arizona were brought to light prior to 1875, and each of them was gophered as soon as possible, the most active and prosperous period for placer mining in the State was from 1858 to about 1880. During that period, prospecting in certain portions of the region was vigorously opposed by the Indians. Before 1885, the cream of the placer gold had been harvested, largely by crude methods of dry-washing and, in some areas, by sluicing, rocking, and panning. In order to rework the gravels for the gold not recovered by the early miners, various attempts at dredging, hydraulicking, and large-scale dry concentration have been made, but so far the majority of these efforts have been unsuccessful. In general, the placer industry of Arizona during the last forty years has been unsteady and has depended upon such factors as unemployment, seasonal rainfall, or promotional enterprises.

PRODUCTION

The total production of Arizona's placers is difficult to estimate. Because the major production was during the early frontier days, when no records were kept, any estimate must be based largely on information secured from pioneers of the various districts. Such figures are necessarily inaccurate. They may be too low, because of the generally intermittent, often scattered operations of transient miners; they may be too high, because of the human tendency to exaggerate regarding gold; or, they may be lacking for districts where no remaining person remembers anything about early production. Furthermore, many of the early-day migratory miners secretly carried their gold in belts or in packs with them when they left the country and spent at the local towns only what was required for supplies and pleasure.

The table given on page 15 is based on very conservative estimates which, for certain districts, may be far too low. In the table shown opposite page 16, the U. S. Mineral Resources[6] yearly production records for 1900-1931 are given. This reported production from 1900-1931, inclusive, was $899,039.

---

[6] Published annually by the U. S. Geological Survey up to and including the year 1923, and by the U. S. Bureau of Mines since 1923.

ESTIMATED PRODUCTION OF PRINCIPAL ARIZONA GOLD
PLACERS PRIOR TO 1900

Field	Estimated production	Source of Estimate
La Paz.	$2,000,000	J. Ross Brown, 1868
Gila City.	500,000	J. B. Tenney
Laguna Muggins.	150,000	
Kofa.	40,000	E. L. Jones, Jr., U. S. G. S. Bull. 620
Castle Dome.	75,000	J. B. Tenney
La Cholla Middle Camp. Oro Fino Plomosa.	300,000	
Weaver. Rich Hill.	2,000,000	J. B. Tenney
Lynx Creek.	1,000,000	W. Lindgren, U. S. G. S. Bull. 782
Hassayampa. Big Bug. Groom Creek. Minnehaha.	1,000,000	J B. Tenney
Greaterville.	700,000	J. B. Tenney
Quijotoa.	250,000	J. B. Tenney
Total	$8,015,000	

YUMA COUNTY

The important gold placer districts of Yuma County are Gila
City, Laguna, Muggins, Castle Dome, Kofa (S. H.), Tank, Trigo,
La Paz, La Cholla, Oro Fino, Middle Camp, Plomosa, and Harquahala. These districts, which are in one of the most arid portions of the Southwest, have but little water outside of the Colorado and Gila rivers. The climate of the region is uncomfortable
for placer mining during the summer, but very enjoyable in
winter. According to the U. S. Weather Bureau, Quartzsite, which
is near the Plomosa, La Paz, and Dome Rock placers at an elevation of 800 feet above sea level, has a mean annual rainfall of 6.53
inches, a mean annual temperature of 69 6°, a maximum temperature of 119°, and a minimum of 9° above zero on record. Likewise, Yuma, which is about twenty miles from the Laguna and
Gila City placers at an elevation of 141 feet, has a mean annual
rainfall of 3.13 inches, a mean annual temperature of 71.7°, a maximum temperature of 118°, and a minimum of 22° above zero.

According to Heikes, Yuma County placer production from 1860 to 1880 is estimated at $20,000,000 or more in gold, but these figures are probably excessive. The 1900-1931 recorded yield was about $144,000.

GILA CITY OR DOME PLACERS

Situation and accessibility: The Gila City placers, which are at the northern end of the Gila Mountains, about twenty miles east of Yuma, have been worked over an east-west length of approximately two miles and a width of from ¼ to ¾ mile. Gila City was about 1½ miles west of the present site of Dome, near the mouth of Monitor Gulch. The Southern Pacific Railway and the old Yuma-Gila Bend road skirt the northern margin of this placer ground.

History: The Gila City Placers became well known in 1858. Hinton,[7] in 1878, recounted their early history as follows: "Within three months of their discovery, over a thousand men were at work prospecting the gulches and canyons in this vicinity.
The earth was turned inside out. Enterprising men hurried to the spot with barrels of whiskey and billiard tables. Jews came with ready-made clothing and fancy wares; traders crowded in with wagonloads of pork and beans. There was everything in Gila City within a few months but a church and a jail. The diggings continued rich for four years and have been continuously worked on a smaller scale up to the present time."

Farish[8] states that Lieutenant Mowry found, in 1859, about 100 men and several families working the gravels at Gila City and saw more than $20 washed from eight shovelfuls of dirt. He was told that from $30 to $125 per day was recovered by each worker.

Although the cream of their production was skimmed before 1865, these placers have been worked more or less every year down to the present time, and all the known productive gravel areas have been dug over at least once.

So far, this gold has been commercially recoverable only by dry washing or by panning of dry-washer concentrates at the river. Many plans have been made for large-scale recovery of the gold, but few of them ever passed the experimental stage. One such enterprise, attempted in 1870, has been mentioned by Raymond[9] as follows: "At Gila City a San Francisco company has during the last year erected works to pump water from the Gila up into a large reservoir on top of the highest foot-hills in

---

[7] Hinton, R. J., Handbook to Arizona. San Francisco, 1878

[8] Farish, T. E., History of Arizona, vol. 1, pp 296-297. 1915.

[9] Raymond, R. W., Statistics of mines and mining in the states and territories west of the Rocky Mountains, (1870), p. 272. Washington, 1872.

Year	Cochise County	Gila County	Greenlee County	Maricopa County	Mohave County	Pima County Greater-Vile	Quijotoa	Total	Pinal County
1900									
1901									
1902									
1903						$ 1,920	$ 5,400	$ 7,320	$ 416
1904						2,500	6,400	8,900	3,229
1905				$ 705		5,033	9,046	14,079	539
1906	$ 1,939			301		2,839	3,014	5,853	281
1907	843	$ 127	$ 415	1,296		5,032	3,810	8,842	446
1908	180		662	1,726		3,109	1,058	4,167	542
1909		68	311	696	1,380	2,209		2,209	
1910	521	215	994	1,109		2,146	251	2,397	
1911	115	276		965		2,067		3,858	1,164
1912				1,356	35	3,557	927	4,484	793
1913			203	1,070		3,495		8,431	1,111
1914	228	146	2,280	1,564		2,393		7,461	506
1915	1,789		813	1,001				2,650	608
1916	1,946		111	414				1,799	225
1917				3,108				4,877	
1918			314			416		416	224
1919	1,267			390					
1920				208		1,226		1,226	
1921	499	527	618	934		1,511		1,621	416
1922	85	243	627	775	751			2,906	230
1923	319		1,413	419	174			622	302
1924				388				211	319
1925				125				126	
1926	113		180	684	190			271	
1927	33		79	123				712	
1928			511	174	173			620	
1929				160		511		511	
1930	154		159	249	557	1,257		1,257	
1931	736	704	410	1,149	193	1,535		1,689	98
Total 1900–1931	$10,767	$ 2,306	$10,050	$21,089	$ 3,453	$42,756	$29,906	$99,515	$11,449

GOLD PLACERS, AS REPORTED BY U. S. MINERAL RESOURCES.

Santa Cruz County	Yavapai County				Yuma County			Total Arizona
	Big Bug	Weaver	Walker	Total	Castle Dome	Plomosa	Total	
								$188,097
								105,024
								43,407
								11,751
				$ 124		$ 4,000	$ 4,000	16,846
		$16,273		26,697		829	829	42,849
		9,466		17,848		4,754	14,463	40,685
		11,011		21,087		3,532	12,072	45,128
$ 299		6,089		13,708	2,737	2,057	9,787	31,071
92		6,980		15,599	2,205	2,136	8,385	28,740
871	$ 2,121	5,993		11,595		2,383	8,426	26,078
622		4,813		9,603		2,498	7,118	23,721
	6,192	4,839		16,001		9,523	20,612	43,281
	2,369	5,515		10,272			9,767	30,854
308	2,038	4,210	3,733	12,238	1,413	2,061	5,542	30,273
509	4,147		4,587	13,977	1,103	3,608	14,058	35,405
	1,692		1,372	3,609	534		6,290	14,394
			2,424	3,780	1,822	1,172	5,636	17,401
			1,647	2,205		907	1,108	4,267
			1,225	2,101			976	4,734
				3,149				4,583
	1,529		1,104	5,891			2,108	12,614
			1,225	5,835		297	642	12,094
			2,356	4,532			1,132	8,913
		310	765	2,240				3,157
	1,506		314	2,360	433	1,103	1,672	4,287
	847	901	445	4,730			874	6,862
135				4,020			1,068	6,280
	484	1,144		3,930		734	1,016	6,424
	175	711	2,764	4,650		354	354	5,675
	3,242	1,854	2,378	8,403		1,453	2,291	13,070
690	4,409	3,866	1,034	12,083		1,425	4,322	22,074
$ 3,526	$30,751	$83,975	$27,373	$242,267	$10,247	$44,826	$144,548	$890,039

order to work the placers of the vicinity by hydraulic power.
They use a nine-inch pipe through which they pump the water."
Numerous gold-saving machines, large and small, have been tried
out here, but most of them were of inadequate design. The remains of one ponderous screw-trommel device, brought here
scores of years ago, are still visible.

The total production of the Gila City placers has been roughly
estimated by J. B. Tenney at $500,000, most of which was made
prior to 1865 Their annual output during the seventies amounted
to a few thousand dollars.[10]

Topography: The Fortuna and Laguna topographic sheets, issued by the U. S. Geological Survey in 1929, include the Gila
City placers Opposite the northern end of the Gila Mountains,
the Gila River bottom lands, which lie about 165 feet above sea
level, are bordered on the south by a gently northward-sloping,
dissected bench that rises abruptly from 35 to 300 feet higher.
From this bench, which is from ¼ to one mile wide, the main
mass of the Gila Mountains rises steeply. Numerous canyon systems, originating in the mountains, have cut steep, northward-
trending gulches, from 35 to 150 feet deep, in this bench

Local geology: Faulted against the schist of the main mountain mass is the series of probable Tertiary sedimentary rocks
that constitute the bedrock of the bench and of the placer deposits. These beds consist of well-stratified, weakly consolidated,
locally mud-cracked clays, marls, arkoses, and sandstones. Their
color is pale gray, buff, light green, or red, and their texture is
generally fine grained, even to the very base of the mountains
This consistently fine-grained character indicates that they were
deposited when no high mountains were very near, and the well-
developed, locally mud-cracked strata point to deposition in shallow water bodies of considerable size.

More or less faulting and tilting are evident throughout this
formation. In the road and railway cuts about 2½ miles north
of Blaisdell, the beds strike N 80° E and dip 25° SE. The age
of the sediments is regarded as probably Tertiary, although, as
Bryan[11] pointed out, they are not as thoroughly cemented as the
Tertiary sediments east of Wellton.

After tilting, these beds were beveled to a pediment Overlying
this pediment and capping the smooth-topped spurs of the dissected bench is a mantle of gravel, up to fifteen feet thick. This
mantle extends across the fault that separates the Tertiary (?)
sediments from the schist, and continues, as narrowing terraces,
for some distance headward into the canyons of the main mountain mass. Most of the material in these gravels appears to rep-

---

[10] Raymond, R. W , Work cited, volumes for 1872-1875.
[11] Bryan, Kirk, U. S. Geol. Survey Water-Supply Paper 499, p. 63 1925.

resent outwash from the Gila Mountains, but part of it is residual from erosion of the Tertiary (?) beds. Bryan[12] interprets the outwash as having been deposited when the Gila River bed stood about 75 feet above its present level. The age of the gravels is regarded as Quaternary.

The gulches that dissect this terrace are floored by gravel, sand, etc., that are partly of local origin, but mostly have been swept down by flood-waters from the mountains. At the edge of the mountains, this material contains subangular to rounded boulders that are as much as two feet in diameter, but, northward, it becomes progressively finer.

Gold-bearing gravels:[13] This Quaternary outwashed material constitutes the gold-bearing gravels of the Gila City placers, and the pediment carved on the underlying Tertiary (?) sediments forms their bedrock. Most of the gold was found at or near bedrock in the gulches, but a considerable amount was recovered from the benches. Practically all the gulches and benches from ¼ mile east to three miles west of Dome carry some gold, but Monitor Gulch, 1½ miles west of Dome, was the scene of the active mining.

Northward from a point not far south of the railway, the bedrock is reported to extend under the water table. Depths of more than fifteen feet to bedrock have not appeared to be profitable for mining.

Origin: The gold of the Gila City placers probably came originally from the various gold-bearing quartz veins in the northern end of the Gila Mountains. As no high-grade veins have yet been found there, the negative conclusion that many pockety or small low-grade veins supplied the gold seems most reasonable. During deposition of the fine-grained Tertiary (?) sediments, the Gila Mountains probably were marked by very low relief, slow erosion, and relatively deep rock decay. After each period of subsequent uplift, they suffered rapid erosion, and the weathered quartz veins of the decayed rocks readily parted with their gold. Floods in the young canyon systems swept this detritus northward, dropping out the gold as the stream gradients lessened. Further milling of these gold-bearing gravels by repeated floods concentrated the gold along the bottom of the channels where the clayey bed rock caught it.

Recent conditions: The gold not yet mined from these gravels is distributed in a rather spotty fashion. In 1926, Messrs. Neal and Morgan found an $88.00 nugget on one of the benches near Monitor Gulch. They found the gravel to run about fifty cents per cubic yard in a few cuts, but ten cents or less in many places.

---

[12] Work cited, p. 67.
[13] Part of this information was furnished the writer by Messrs. Robert Morgan, Harry McPhaul, and the late W. M Neal.

The fineness of this gold was about $19.00 per ounce.[14] About half of the nuggets were larger than match heads, and a fourth of them were from $3 to $6 in value. Almost all of the gold particles were rough, and the $88 nugget contained some white quartz.

During part of 1931, Mr. G. H. Mears attempted small-scale hydraulicking operations in Monitor Gulch. Water for this enterprise was obtained from a shallow well near the railway and pumped through about ¼ mile of small pipe.

During the cool portion of the 1932-1933 season, approximately 25 men, mostly transients, were conducting dry placer operations in the Gila City area. The daily earnings per man ranged from a few cents up to generally less than $1.00. During 1932, Mr. E. H. Rhodes, storekeeper at Dome, purchased $2,296 worth of gold which came partly from the Gila City and partly from the Muggins placers.[15] Besides this amount, an unknown quantity from these areas was marketed elsewhere.

LAGUNA PLACERS

The Laguna or San Pablo Mountains, which are in Ranges 21 and 22 W., immediately north of the Gila River and the Gila Mountains, contain gold placers in their southern, southeastern, and southwestern portions.

The Laguna quadrangle sheet, issued by the U. S. Geological Survey in 1929, shows the topography of this range.

McPhaul area: Considerable placer mining has been done in the southern portion of the Laguna Mountains, from near the Gila River to about 1¼ miles north of McPhaul Bridge. Most of this work was done many years ago, but a little dry-washing is still carried on. Only scanty production records for this particular area are available. During some years, its yield was lumped with that of Gila City.

These placers, which conform to the exposure of tilted, beveled, Tertiary (?) sediments that constitute their bed rock, occupy an area of approximately ¾ square mile, limited on the north and east by the hard rocks of the Laguna Mountains, on the south by the Gila River bottom lands, and on the west by the high gravel capping of the range. The Tertiary (?) strata, whose general character has been described on page 17, strike and dip in various directions, but a northerly or northwesterly strike appears to predominate. Many southeastward-trending arroyos have dissected the area. Most of the evidences of placer mining activity are confined to the inter-arroyo benches near the base of the overlying gravels, but some at lower elevations and also along the arroyo bottoms are evident.

----

[14] Oral communication
[15] Oral communication from Mr. Rhodes.

Las Flores area: Las Flores district, in the southeastern portion of the Laguna Mountains and 1¼ miles north of the Gila River, is accessible by ¾ mile of road that branches northwestward from the Yuma-Quartzsite highway at a point about 3¼ miles from McPhaul Bridge.   It is near the head of an alluvium floored gulch, at an elevation of 300 to 400 feet above sea level. The erosion of several gold-bearing quartz veins in this district has given rise to small placer deposits.

According to Raymond,[16] placer mining was carried on in Las Flores area chiefly by Mexicans and Indians, at about the time when the Gila City placers were most active.   Part of this placer gold occurred in the vicinity of the Golden Queen and India claims, and some was followed downstream to the bank of the Gila River.   A little placer mining has been done in several gulches along the southern margin of the mountains.   No record or estimate of the amount of gold recovered is available.

Laguna Dam area: At the eastern end of Laguna Dam, about ten miles northeast of Yuma, masses of black schist and coarse, granitic gneiss rise steeply for 250 feet above the Colorado River. Erosion of certain quartz veins in these rocks has given rise to coarse, rusty placer gold that, in places, extends into the bed of the Colorado River.   In 1884 or 1885, a crude attempt was made to recover this river-channel gold by dredging, but a flood destroyed the dredge before it attained any success.

In 1907, during the construction of Laguna Dam, placer nuggets and a small gold-quartz vein were found[17] at the river margin of these mountains.   Considerable placer prospecting has been done in several of the gulches of this area, and certain pot holes, up to 100 feet above the river, were found to carry rather coarse gold.   This coarseness points to a local origin rather than to a long transportation by the Colorado River.   The U. S. Mineral Resources report from the Laguna placers a production of $1,457 in 1910 and $1,989 in 1912.   The pot holes yielded most of this amount and have made some production since then.

Similar, but more extensive, pot-hole placers occur on the California side of the Colorado River.

Recent operations: During the cool portion of the 1932-1933 season, a maximum of fifty men were conducting dry-placer operations in the McPhaul and Las Flores areas.   All of this ground is privately owned, but, in general, no royalties were charged.   According to Mr. Harry McPhaul,[18] the average daily earnings per man were between 50 and 75 cents.   At the same time, approximately 25 men were placering in the Laguna Dam area.

---

[16] Raymond, R W , Statistics of mines and mining in the states and territories west of the Rocky Mountains (1870), p. 272. Washington, 1872.

[17] Oral communication from Mr. A B. Ming

[18] Oral communication.

MUGGINS PLACERS

The Muggins Mountains, which occupy parts of Twps. 7 and 8 S., R. 8, 9, and 10 W., contain gold placers in their southern and central portions. These placers have been known for many years but, because of being less easily accessible than the neighboring Gila City area, they have not been so intensively worked.

The Wellton, Fortuna, and Laguna quadrangle sheets, issued by the U. S. Geological Survey in 1929, show the topography of part of the Muggins Range.

Southern portion: In the southern portion of the range, the major placers occur in Burro Canyon. Minor ones are found in certain smaller conyans in the vicinity of a prominent mountain that is variously known as Klotho, Coronation, or Muggins Peak, and also at the southern base of Long Mountain. Burro Canyon, which is accessible from Dome by some ten miles of unimproved road, trends southward from Muggins Peak. Here, southward-dipping lava flows, intercalated with thick beds of conglomerate, form a rugged terrain. This conglomerate, which consists mainly of coarse, subangular pebbles of gneiss and granite, rather firmly cemented in a sandy to clayey matrix, forms the bedrock of the placers. The gold-bearing gravels occur principally as ancient bars several feet above the stream channel and, to a less extent, in the present stream bed. The gold occurs as particles up to 0.15 inch in diameter, mostly concentrated at or near bedrock. It appears to have been derived by erosion of the conglomerate, in which it was probably present as low-grade placer material derived from gold-bearing quartz veins originally contained in the gneisses, schists, and granites of the range.

Central portion: Gold placers occur in the central portion of the Muggins Mountains, in the vicinity of the headward forks of the long, northwestward-trending canyon that bisects the range. The gravels of this canyon, which are reported to have yielded many rich pockets during the early days, are occasionally worked after heavy rains This gold probably accumulated from the disintegration of quartz veins contained in the adjacent schists and gneisses.

Recent operations: During the cool portion of the 1932-1933 season, a maximum of approximately 25 men were working in the southern Muggins placer area. The daily recovery per man was generally less than $1 00. Practically all the water used must be hauled from wells in the Gila Valley. During 1932, Mr. E. H. Rhodes, storekeeper at Dome, purchased $2,296 worth of gold which came partly from the Muggins and partly from the Gila City placers.[10] Besides this amount, an unknown quantity from these areas was marketed elsewhere.

---

[10] Oral communication from Mr Rhodes

CASTLE DOME PLACERS

The major gold placers of the Castle Dome Mountains are east and south of the Big Eye mine, which is 31 miles by road northeast of Dome. The gold, which occurs mostly at or near bedrock in gulches, appears to have been derived from the erosion of gold-bearing veins in the vicinity.

These placers were discovered in 1884, but their production to the end of 1902 is unknown. The U. S. Mint report for 1887 states that the field was being worked in a crude way by Mexican dry-washers. According to data compiled by J. B. Tenney, of the Arizona Bureau of Mines, the production rate from 1908 to 1916 was about $3,000 per year, and since 1908, a total of $25,-000 has been reported. Assuming this rate from 1884 to 1908, the yield from these placers for that period would amount to between $75,000 and $100,000.

Recent operations: During 1932-1933, seldom more than two or three men were working in this placer field. Operations are hampered by the scarcity of water in the region.

KOFA OR S. H. PLACERS

A small area of gold placers situated in the Kofa or S. H.

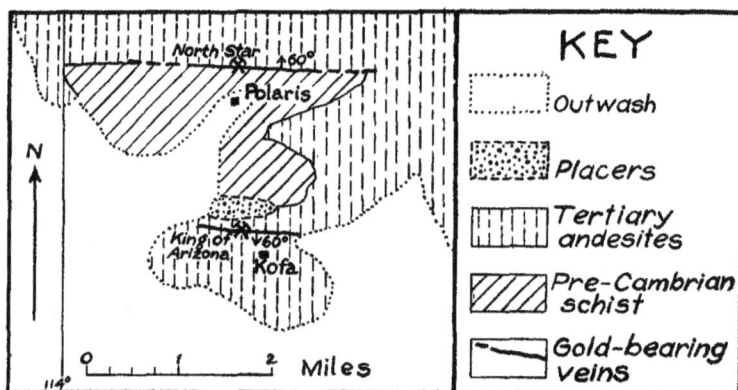

Figure 3. Preliminary geologic reconnaissance map of the Kofa or S. H. placer region, after E. L. Jones, Jr. and N. H. Darton.

Mountains of central Yuma County, about 56 miles northeast of Yuma, has been described by Jones.[20] A geologic sketch map of the vicinity is shown in Figure 3. Of these placers Jones says:

"The known placer deposits of the Kofa Mountains occur in a gulch draining westward north of the detached hills in which the King of Arizona Mine is located. These placers have been

[20] Jones, E. L., Jr., A Reconnaissance in the Kofa Mountains, Arizona: U. S. Geol. Survey Bull. 620, p. 164. 1916.

worked for many years, and the production is reported to be about $40,000 in gold nuggets. At present (1914) the placers are being worked in a small way, and a yearly production of several hundred dollars is reported. The gold occurs in outwash deposits which consist of boulders and fragments from the metamorphic and volcanic rocks. The gold-bearing debris is said to be from a few feet to seventy feet deep over an area of approximately sixty acres. The gold is coarse and occurs near bedrock. It has evidently been derived from the disintegration of auriferous veins in the metamorphic rocks, as it is much coarser than that contained in the North Star and King of Arizona veins."

Recent operations: During the winter and spring of 1932-1933, eight to ten men were working in the Kofa placer area. The average daily recovery per man was 75 cents or less.

TANK MOUNTAINS PLACERS

Some placer gold has been mined in the Tank Mountains at various times since the seventies, but no record or estimate of the production exists.

Probably the earliest and most profitable activity was in the main gulch below the Johnnie or Engesser prospect, in the northwestern portion of the range. This placer gold was probably derived from local gold-bearing veins. As the field was small, its richer ground was soon worked out, but, during the past few years, it has occasionally produced a small amount of gold.

Some thirty years ago, active dry-washing was carried on in certain shallow bench and stream gravels on the pediment near the Puzzles, Golden Harp, Ramey, and Regal prospects, which are at the eastern foot of the range The gold obtained from the Puzzles area is said by Mr. John Collins[21] to have been coarser than that from the other localities.

Recent production from all of these areas has been practically negligible.

TRIGO PLACERS

The Trigo placers are at the western base of the Dome Rock Mountains, in T. 2 N., R. 21 W, and approximately 22 miles by road from Quartzsite. The gold-bearing gravels occur in certain arroyo bottoms and in ancient bars and channels. Most of the gold is in the form of flat grains that are worth approximately $19.40 per oz.[22] For many years, small-scale, intermittent dry-washing operations have been carried on in this field, but no record or estimate of the production exists. Operations are greatly hampered by the scarcity of water and the cemented character of the gravels.

---

[21] Oral communication.

[22] Written communication from Mr. W G Keiser, of Quartzite.

La Paz Placers

Situation and accessibility: The La Paz placers are south of the Colorado River Indian Reservation of west-central Yuma County, along the western foot of the Dome Rock Mountains, about nine miles west of Quartzsite and six miles east of the Colorado River (see Figure 4). The district is accessible by some five miles of unimproved road that branches northward from the Quartzsite-Blythe highway.

Topography: The Dome Rock Mountains rise steeply to approximately 2,900 feet above sea level, or more than 2,000 feet above the adjacent plains, and are extensively dissected by deep canyons. From their western foot, a wide dissected bench slopes gently westward to the low bluffs that limit the Colorado River bottom lands. No perennial streams flow through the placer district, but branching arroyos drain the run-off of the rainy seasons to the Colorado River. Most of the water used for domestic purposes is hauled from Quartzsite or from shallow wells near the river. A scanty supply is afforded by Gonzales Wells, or by uncertain natural rock tanks, such as Goodman Tank

History: According to former State Historian Hall,[23] the presence of placer gold near the Colorado River was learned from the Indians soon after the establishment of the military post at Yuma. These Indians gave a few small nuggets of the gold to a trapper, Capt. Pauline Weaver, and, about 1862, according to Browne,[24] guided Weaver and his party to the rich gravels. The party picked up about $8,000 in nuggets, returned to Yuma for supplies, and spread news of the discovery. Several hundred miners soon rushed to the district, found the placers to be very rich, and established the adobe town of La Paz about 2¼ miles from the river. This town, which soon attained a cosmopolitan population of over 1,500, became a station on the Overland Trail from San Bernardino to Ft. Whipple and was the county seat until 1871.[25] The district flourished until about 1864, when apparent exhaustion of the higher-grade placers and discoveries of new diggings caused a decline in activity. In 1873, 1874, and 1876, additions to the Colorado River Indian Reservation included much of the placer ground and greatly restricted mining. La Paz became practically deserted, and the site of this once flourishing town is now marked only by adobe ruins.

After the area was excluded from the Indian Reservation in 1910, the New La Paz Gold Mining Company acquired control of a large portion of the placer ground and made preparations for large-scale hydraulic treatment of the gravels. Four shallow

[23] Hall, Sharlot M , Personal communication.
[24] Browne, J Ross, Resources of the states and territories west of the Rocky Mountains 1868.
[25] Bancroft, H H , A history of Arizona and New Mexico, p. 616.

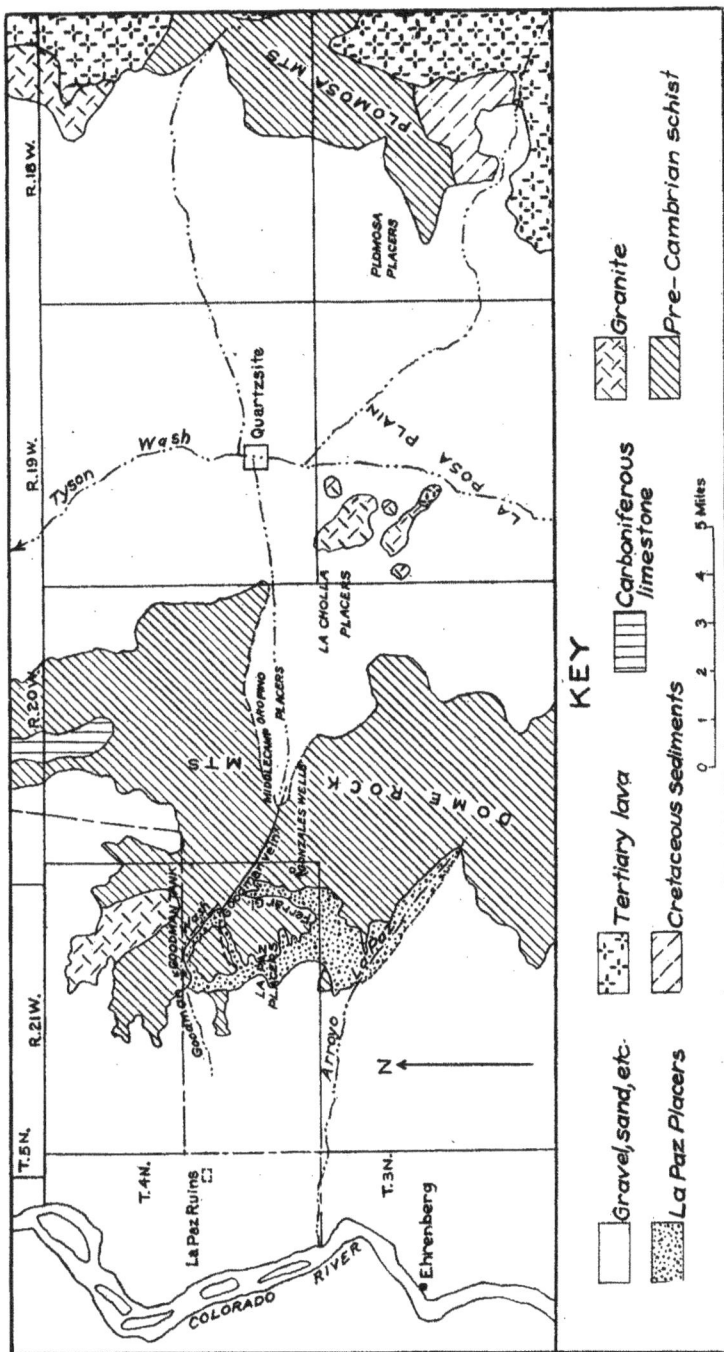

Figure 4. Preliminary geologic reconnaissance map of part of the Dome Rock and Plomosa mountains, showing general location of La Paz, Plomosa, La Cholla, Oro Fino, and Middle Camp placers.

wells were drilled in the Colorado River flood plain some 4½ miles west of the placers. Water was to be pumped from these wells through a twelve-inch pipe line to a reservoir 540 feet above the river, or 225 feet above the placers. Part of the pipe line was installed, but, from 1912 to 1915, the land was again included in the Reservation, and the project was not completed. Several other plans for large-scale operations have been outlined, but none of them have been carried out.

Production: Information on the earlier production of the La Paz placers is given by Browne,[26] who quotes a letter from Mr. A. McKay, a member of the Territorial legislature from La Paz, as follows:

"Of the yield of these placers, anything like an approximation to the average daily amount of what was taken out per man would only be guess work. Hundreds of dollars per day to the man was common, and now and again a thousand or more a day. Don Juan Ferra took one nugget from his claim that weighed forty-seven ounces and six dollars. Another party found a chispa weighing twenty-seven ounces. Many others found pieces of from one to two ounces up to twenty, and yet it is contended that the greater proportion of the larger nuggets were never shown. .... It is the opinion of those most conversant with the first working of these placers that much the greater proportion of the gold taken out was in nuggets weighing from one dollar up to the size mentioned above. .... As has been said above, the gold was large and generally clear of foreign substances. .... All that was sold or taken here went for $16 to $17 an ounce. Since the year 1864 until the present, there have been at various times many men at work in these placers, numbering in the winter months hundreds, but in the summer months not exceeding 75 to 100; all seem to do sufficiently well not to be willing to work for the wages of the country, which are and have been for some time from $30 to $65 per month and maintenance. No inconsiderable amount comes in from these placers now weekly, and only a few days ago I saw, myself, a nugget which weighed $40, clear and pure from foreign substance. ....

"Of the total amount of gold taken from these mines, I am at loss to say what it has been .... I have failed to find any pioneer whose opinion is that less than $1,000,000 were taken from these diggings within the first year, and in all probability as much was taken out in the following years."

According to Hall,[27] local gold nuggets and dust were the principal currency, particularly for gambling, in La Paz; but a large portion of the gold obtained by the Mexican placer miners went to Mexico.

On account of the crude methods of recovering the gold, en-

---

[26] Browne, J. Ross, Work cited, 1868.
[27] Personal communication.

tirely by dry washing in pans or wooden "bateas," it is apparent that only the coarser gold could be saved, and only extremely rich ground would be payable. Wet methods were out of the question, for, according to Jones,[28] water packed from the town of La Paz to the placers brought $5 a gallon during the rush period. With the introduction of dry-washer machines in the late sixties, greater quantities of material could be handled and a greater percentage of recovery effected, but, by that time, most of the richer ground had been worked over.

Local geology: The Dome Rock Mountains in this vicinity consist largely of metamorphic rocks and granite (see Figure 4). For a short distance west of the foot of the range, these rocks floor a dissected pediment and constitute the bedrock of the principal placers. Westward, they disappear beneath extensive deposits of sand, gravel, and clay, which in turn are locally overlain by coarse outwash gravels and boulders.

Distribution and character of the placer gravels: The placers occur mainly in Goodman Arroyo and Arroyo La Paz, and in certain tributary gulches such as Ferrar, Garcia, and Ravenna. According to Jones,[29] "Ferrar Gulch, tributary to Arroyo La Paz, contained the richest and most productive placers of the district. Evidences of former work are seen in the old excavations and . . . . in exposures of bedrock where the wash was shallow. . . . The thickness of the gold-bearing wash is variable, ranging from a few feet on the mountain slopes to an unknown measure in Arroyo La Paz and in the gulch traversed by the (old) Quartzsite-Ehrenberg road. Shafts have been sunk in the wash to depths of thirty feet without reaching bedrock and it is reported that in places the wash is at least sixty feet deep. By far the greater part of the auriferous material is unworked, especially that in the lower courses of the arroyos, where the wash is deep. Ferrar Gulch for most of its course has been practically worked out.

"The gold-bearing material consists of sand and clay inclosing angular rock fragments of greatly variable size. Tests indicate that about twenty per cent of the wash will pass through a quarter-inch screen, and the largest boulders weigh several hundred pounds. The material near the surface is unassorted and is unconsolidated, being readily worked with pick and shovel. That at depths of fifteen or twenty feet is consolidated, but the cementing substances readily disintegrate on exposure to air. Deposits of wash below the depths of test pits may prove to be similar to the outwash on the east slope of the Dome Rock Mountains and in the Plomosa placers, where the material is firmly cemented with calcium carbonate and requires crushing in order to free the gold. The ground stands sufficiently well to permit

---

[28] Jones, E. L., Jr , Gold deposits near Quartzite, Arizona: U S Geol. Survey Bul 620, p. 49. 1916.
[29] Jones, E. L., Jr., work cited, p. 51

the sinking of shafts without the use of timber. The wash is readily worked in dry washer machines, the only requirement being that the ground must be dry. The gold is said to be distributed throughout the wash, though in the early workings the richest yield was obtained near bedrock.

"No estimate could be made of the probable gold content of the wash in the La Paz district because of lack of detailed data and of uncertainty as to the limits of the wash, but in one area the deposit, said to contain values of 50 to 75 cents per yard and much of it thirty feet or more deep, occupies at least 640 acres, and considerable areas extend into the smaller gulches.

Character of the gold: "The size of the gold now recovered from the deposits of the La Paz district probably averages only a few cents, but as already stated, the gold recovered from the early workings was much coarser The gold is rough and angular, and particles of iron cling to some of the nuggets. Magnetite is always found in the concentrates, and boulders of magnetite, the largest weighing several pounds, are frequently found on the surface."

Heikes[30] states that the largest nugget found in this region was valued at $1,150 and assayed about 870 in fineness. Most of the gold particles or nuggets ranged in value from five cents to $10, although $20 and $40 nuggets were not uncommon.

Origin: The La Paz placers were probably derived by the erosion of many gold-bearing veins in the Dome Rock Mountains.

Recent operations: During the winter of 1932-33, from fifty to sixty men were reported to be conducting small-scale, individual dry-washing operations in the La Paz district. The average daily recovery per man was from 50 to 75 cents.

In 1933, all of the claims belonging to the New La Paz Gold Mining Company in T. 4 N., R. 21 W. were under a twenty-year lease to the Great Western Development Corporation of Arizona, Ltd.

PLOMOSA DISTRICT

The Plomosa placer district includes the eastern and western margins of La Posa Plain. This plain, which separates the Plomosa Mountains on the east from the Dome Rock Mountains on the west, is approximately ten miles wide and from 1,000 to 1,300 feet in elevation. It is dissected, particularly in the marginal portions, by many shallow arroyos that are tributary to its northward-flowing axial channel, Tyson Wash. These arroyos contain no water except for short periods after occasional heavy rains. Practically all of the water used in the western part of the district is hauled from shallow wells at Quartzsite.

---

[30] Heikes, V. C., Dry placers in Arizona: U. S. Geol. Survey Mineral Resources for 1912, Part I, p. 259.

Heikes[31] states: "Surrounding the post office of Quartzsite, in the Plomosa mining district, and extending in every direction, covering an area of about 7,500 acres, is found dry-placer ground with values to an average depth of fifteen feet and varying from five to fifty feet. The gold content per cubic yard is reported to average in coarse gold from ten cents to several dollars."

The most important placer fields in the Plomosa district are La Cholla, Oro Fino, and Middle Camp, which lie near the Dome Rock Mountains; and the Plomosa, near the Plomosa Mountains (see Figure 4). These areas have been worked intermittently by individual dry-washers since the early sixties. Several large-scale operations have been planned or attempted, but none have been successfully carried out. The 1900-1931 yield of the Plomosa placer district is given by the U. S. Mineral Resources as $44,826.

During part of the winter of 1932-1933, more than 100 men were reported to be placer mining in this district.

La Cholla Placers

La Cholla placers comprise an area some four or five miles long and of irregular width, bordering the eastern foot of the Dome Rock Mountains south of the Quartzsite-Blythe Highway.

Here, a gently eastward-sloping pediment or rock floor, eroded largely on tilted bluish-gray slates, borders the mountains and, extending beneath the gravels of the plain, constitutes the bedrock of the placers

The gravels in general consist of an ill-assorted aggregate of subangular to slightly rounded slate, schist, and quartzsite fragments, more or less firmly cemented with lime carbonate. They are commonly of medium texture, but range in size from rather fine material to boulders three or four feet in diameter.

The gold occurs mostly at or near bedrock, but some is erratically distributed throughout the gravels Its particles are characteristically angular and crystallized and range in diameter from that of a pin point up to $\frac{1}{8}$ inch or more It has not been transported far and probably was derived from numerous small gold-bearing veins in the adjacent mountains. Black sand is abundant only in the shallower diggings.

Recent operations: During the first half of 1933, the principal activity in La Cholla placers was on a group of three twenty-acre claims held by Messrs G W. McMillen and Guy Hendrix On part of this ground, at the eastern foot of a low, steep spur, many old pits, shallow shafts, and drifts proclaim earlier placer mining activity. A few hundred feet farther east, the present operators sank a shaft that struck bedrock at a depth of 84 feet. According to Mr. McMillen, a small pay streak was cut at a depth of 42 feet, and rather fine gold was found at a distance of fifteen

---

[31] Heikes, V. C., work cited, p. 258.

feet above bedrock. At bedrock, the shaft encountered a rich southeatward-trending channel. When visited in June, 1933, this channel had been followed by some 300 feet of drifts and a minor amount of stopes, but its width and length had not been determined. As shown by these workings, the bedrock surface slopes about 15° southeastward and, due to its structure, forms natural riffles. The richest gold-bearing gravel occurs within a few inches of the bedrock and is particularly concentrated in the vicinity of reefs and undulations on the bedrock surface, or where boulders are abundant.[32] In places, it contains up to an ounce or more of gold per cubic yard. Locally, crevices in the bedrock contain gold for depths of 1½ to 2 feet.

Although mine openings in these cemented gravels require little or no timber, the material mined does not require crushing. It is run through a ¾ inch trommel screen and then conveyed to a bin from which it is passed over a two-tier dry-washer driven by a small gasoline engine. The tailings from this operation contain approximately fifty cents in gold per cubic yard. Production during the first six months of 1933 amounted to about $6,000 in gold that ranged from 920 to 924 fine.[33] Five men are employed.

In June, 1933, the gravels in a secondary surface channel, a few hundred feet north of this shaft, were being mined with a power shovel. These gravels, which are reported to run 75 cents per cubic yard, were being treated experimentally in a wet jig for which water was hauled from a well 3½ miles distant.

Oro Fino Placers

The Oro Fino placers are at the eastern foot of the Dome Rock Mountains, in the vicinity of the Quartzsite-Blythe highway. Here, tilted, beveled shales constitute the bedrock. The gravels, which are relatively thin near the mountains, contain much slaty material. During the early days, these placers were actively mined, but at present are worked only in a small way by individuals. According to Jones,[34] 640 acres of this tract were sampled by the Catalina Gold Mining Company with test-holes, up to thirty feet deep, sunk every few hundred feet. From these samples it was found that the gold content ranged from a few cents to over $1 per cubic yard and averaged 38 cents per yard. The colors ran from less than one cent to 24 cents each, and the gold was of about $19 per ounce fineness. Here the gold-bearing material consisted of unconsolidated rock debris, up to twelve feet thick, and an underlying cemented gravel eighteen or more feet thick.

---

[32] Oral communication from Mr. McMillen.

[33] Oral communication from Mr. McMillen.

[34] Jones, E. L., Jr., Gold deposits near Quartzite, Arizona: U. S. Geol. Survey Bul. 620, p. 49. 1916.

MIDDLE CAMP PLACERS

The Middle Camp placers comprise an area four or five miles long from east to west by a mile wide at the eastern foot of the Dome Rock Mountains, immediately north of the Oro Fino placers. Here, according to Church,[35] "rich seams of gravel on bedrock yield from four to ten times the value of the thicker gravels, and in crevices there have been found nuggets worth $10 to $25."

Recent operations: During 1932, two companies attempted large-scale operations in this tract.

On ground leased from Middle Camp Placer Gold, Inc., La Cholla Mining Co., Ltd., tried out a large machine equipped with a 3½-yard dragline shovel, approximately 100 feet of sluice boxes, and settling tanks for water recovery. This machine, for which water was hauled from Quartzsite, operated for only a few weeks.

The American Coarse Gold Corporation installed a plant equipped with a dragline shovel and two Cottrell tables. It was operated, with water hauled from Quartzsite, for about two weeks.

In June, 1933, approximately twenty individuals were carrying on small-scale dry-washing in the Middle Camp placers.

PLOMOSA PLACERS

The Plomosa placers are at the estern edge of La Posa Plain and the western foot of the Plomosa Mountains, about five miles southeast of Quartzsite (see Figure 4). This field produced considerable gold soon after its discovery in the early sixties, but no record of the amount is available. Bancroft,[36] in 1909, found that portions of the ground had been honeycombed with small tunnels. Various attempts have been made to work these placers on a large scale, both by dry and wet methods, but, so far as is known, all of the production has been made by individual, small-scale dry-washing.

Geology: The general geology of the Plomosa placer district is indicated in Figure 4. The Plomosa Mountains, which, east of the district, are about two thousand feet above sea level, or one thousand feet above the plain, consist largely of schist, granite, and later volcanic rocks. The schists, which contain gold-bearing quartz veins and stringers, were probably the original source of the Plomosa placers.

According to Bancroft,[37] the placer gravels, which occur in certain old drainage channels leading away from the southwestern

---

[35] Church, John A., in Heikes, V. C., work cited, p. 258.
[36] Bancroft, H., Reconnaissance of the ore deposits in northern Yuma County, Arizona: U. S. Geol. Survey Bul. 451, p. 88. 1911.
[37] Bancroft, H , work cited.

part of the mountains, are made up of fragments of schist, granite, and quartz, cemented by lime carbonate. This conglomerate or "cement rock" varies in thickness from a few inches up to many feet, depending largely on the shape and size of the former channels, and rests upon grayish-green, schistose bedrock.

Regarding the placers, Heikes[38] quotes extracts from a professional report by John A. Church as follows:

"In some localities pits have been sunk to a depth of twenty, thirty, and fifty feet or more to beds of cement which are richer than the gravel. Near the mountain the gold is coarser, but the gravel is much less. Miles of the great deposit, extending westward from the mountains and from three to four miles in width, have been cut into by deep ravines, and they afford miles of banks ten to fifteen feet high in which the upper layer of gravel is well exposed. From these banks, as far as investigations could be made, samples gave an average return value of 64 cents per cubic yard with gold estimated at $18 an ounce. . . . . There were no failures. The results lay between the extremes of 42 cents and $1.04 per cubic yard. The limit of the gravel actually explored was 2,400 by 1,500 feet and eight yards deep . . . . Within this area bedrock was not reached at any time."

Recent operations: The known favorable ground in the Plomosa placer district is held by private locators who carry on intermittent small-scale dry-washing operations each year. During the winter of 1932-33, approximately twelve men were thus engaged there. The average daily earnings per man were reported to be from 25 to 50 cents.

HARQUAHALA PLACER

Mr. L. C Shattuck,[39] of Bisbee, states that, in 1886 and 1887, he worked a small placer in Harquahala Gulch, which is in the southwestern portion of the Harquahala Mountains, eight miles south of Salome. For a short while, Shattuck and his partner each recovered about an ounce of gold per day. Although long since worked out, this placer is of interest because of occurring in the immediate vicinity of the rich Harquahala or Bonanza lode, which was not discovered until 1888.

YAVAPAI COUNTY

The principal gold placers of Yavapai County are in the Lynx Creek, Weaver, Rich Hill, Copper Basin, Big Bug, Hassayampa, Minnehaha, Groom Creek, Placerita, Model, Black Canyon, Granite Creek, Eureka, and Humbug regions.

---

[38] Heikes, V. C., work cited, pp. 257-258.
[39] Oral communication.

LYNX CREEK PLACERS

The Lynx Creek placers are in central Yavapai County, along Lynx Creek from near Walker, seven miles southeast of Prescott, to its junction with Agua Fria Creek, thirteen miles east of Prescott.

Lynx Creek, which flows northward between foothill ridges of the Bradshaw Mountains, and northeast and eastward through the conglomerate terraces of Lonesome Valley, has an approximate length of eighteen miles. Since it extends between elevations of about 7,000 and 4,600 feet above sea level and drains a large, high region, it receives a considerable amount of water each season and is perennial in its upper, pine-wooded course. At Prescott, which is about five miles west of the creek at an elevation of 5,320 feet above sea level, the normal annual fall of rain and snow water is 18.52 inches, the highest temperature recorded was 105°, and the lowest 12° below zero.[40]

History and production: According to former State Historian Hall,[41] the Lynx Creek placers were discovered in 1863 by a party of California miners headed by Capt. Joe Walker. As the news of their discovery filtered back to California, the number of placer miners on Lynx Creek increased to 200 or more. Active work, with hand rockers, pans, and small sluices, continued along the stream for several years before the exhaustion of the richest gravels.

Like most of the placers of the Southwest, unfortunately, no records of the early-day yield are available, but Lynx Creek is noted as one of the most productive gold-bearing streams in Arizona. Raymond[42] reported its 1874 production at $10,000, and Hamilton[43] estimated the total prior to 1881 at $1,000,000. According to Mr. A C. Gilmore,[44] of Prescott, about 100 men were working the Lynx Creek placers prior to 1885, and some of them recovered about $20 per day. Mr. W. R. Shananfelt, of Prescott, states that one man recovered $3,600 in eleven days from the lower reaches of the creek. As shown in the table opposite page 16, the production recorded from 1914 to 1931, inclusive, was $27,373.

Much money has been spent in efforts to work these placers on a large scale. In the late eighties, an Englishman, B. T. Barlow-Massick, built a small dam above the present Prescott-Dewey highway bridge, installed a few miles of thirty-inch pipe, and did some hydraulicking, but ·a flood destroyed the dam. About

---

[40] Smith, H. V , Climate of Arizona: Univ. of Ariz College of Agriculture Bul 130, 1930.

[41] Personal communication

[42] Raymond, R. W , Statistics of mines and mining in the states and territories west of the Rocky mountains for 1874.

[43] Hamilton, Patrick, Resources of Arizona. 1881.

[44] Oral communication.

1900, the Speck Company tried out an old dredge a short distance below the bridge, but the roughness of the bedrock there prevented its success. Later, Mr. G. S. Fitzmaurice operated this

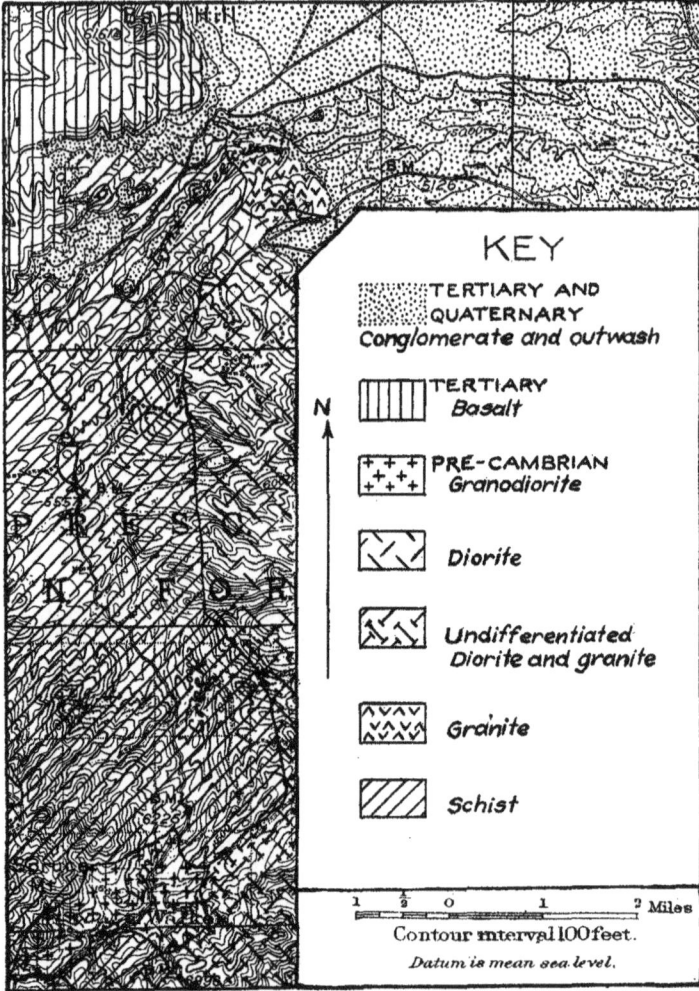

Figure 5. Geologic map of the Lynx Creek region.

dredge farther down the creek, but, after recovering about $800 worth of gold, the dredge fell apart. A large, expensive, patented gold-saving machine was tried out nearby at about this time, but also without success.

In 1927, the Lynx Creek Mining Company attempted large-scale operations with a moveable plant consisting of an Insley excavator, a Barber Green stacker, screens, and sluices. A large yardage of material was treated, but apparently the enterprise did not succeed.

Geology: The geology of the Lynx Creek placer region is indicated on the accompanying map (Figure 5). The oldest rocks are coarse to mediumly fissile schists of sedimentary and igneous origin, extensively intruded by slightly schistose dikes of granite, pegmatite, and diorite. These schists strike approximately N.-S. and dip steeply. Larger masses of dark hornblende diorite and light-colored medium-grained granite intrude these schists. In the northern portion of this area, these pre-Cambrian rocks are overlain by a conglomerate of medium-grained, fairly well rounded gravels, firmly cemented in a matrix of sand and volcanic ash. This conglomerate, which constitutes the bedrock of the placers of lower Lynx Creek, appears to be overlain by the late Tertiary basalts in Bald Hill. The youngest formation in the region is the series of gravels, sand, and boulders that occupy the bed and flood plain of Lynx Creek. This material, which contains the placer gold, is generally well-rounded, except in the upper reaches of the stream.

From near Walker to a point about eight miles in air line downstream, or to the Lynx Creek Mining Company dam, two miles below the bridge, the placers occur as thin benches or bars whose few yards of width can not be shown on a map the scale of Figure 5. Downstream from that point, in the bottom of the steep-walled gulch formed in the conglomerate fill of Lonesome Valley (see Plate 1), the placers attain a maximum width of over one-eighth mile and a thickness of eight to twenty-four feet. It is said that, although some gold is present throughout this eight to twenty-four feet of thickness, the richest material is at the conglomerate bedrock and in a four-foot streak about two feet higher. Lindgren[45] states that the average value is reported at eighteen cents per cubic yard.

According to Lindgren, "At Walker the placers yielded nuggets worth as much as $80, at about $16 an ounce. Lower Lynx Creek produced a finer-grained gold of higher value, worth about $18 an ounce. Such an enrichment in the value of the gold is common and indicates a solution of the silver by the waters." The gold of lower Lynx Creek ranges from finely divided material up to $4- or $5-nuggets, and is associated with considerable hematitic and magnetitic black sand.

The placer gold of Lynx Creek was doubtless derived from disintegration of the numerous gold-bearing quartz veins contained in the pre-Cambrian rocks of the Walker region.

Recent dredging operations:[46] During 1932, a California type

---

[45] Lindgren, Waldemar, Ore deposits of the Jerome and Bradshaw Mountains quadrangles, Arizona: U. S. Geol Bul. 782, p. 109. 1926.

[46] Most of this information was furnished by Mr. H C Sellers, Superintendent of the Calari Dredging Corporation.

Plate 1. Calari Dredging Company operations on lower Lynx Creek in June, 1933.

dredge was installed in the lower Lynx Creek placer area, on the G. S. Fitzmaurice property, below the dam illustrated in Plate 2. Dredging was begun in March, 1933, by the Calari Dredging Corporation, and, from April 1 to June 1, 60,000 cubic yards of gravel, which yielded approximately 32 cents per cubic yard, were treated.

Plate 2.   Dam and reservoir on lower Lynx Creek.

The dredge (see Plate 1) is fifty feet long by 35 feet wide by three stories high and has a capacity of 100 cubic yards per hour. It draws thirty inches of water and normally requires about 85 gallons of new water per cubic yard of gravel treated. Approximately twenty men were employed to conduct the operation three shifts per day. The dredge crew is three men, and the dragline crew one man, per shift.

In June, 1933, the dredging was being carried on to an approximate depth of six feet. The gravel, as mined with a 1½-yard dragline shovel, was passed through a ten-inch grizzly, then through a trommel with a 5-16 inch screen, whence the oversize went to a stacker, and the undersize into a sluice equipped with 400 square feet of angle-bar riffles.

Of the total gold present in the gravels, from 85 to 90 per cent is extracted. It ranges in size from flour up to fragments 0.1 inch in diameter and is accompanied by abundant black magnetitic sand.

Recent small-scale operations: During the spring and summer of 1933, approximately thirty men were rocking and sluicing on upper Lynx Creek. Most of the gravel was obtained in small, dry side gulches and packed to water. In places, trees were being uprooted in order to reach the pay dirt beneath them. A short distance below the Dewey highway bridge, one man was drifting on old side-gulch channels.

According to Mr. A. S. Konselman,[47] of Prescott, who has kept accurate records of the gold produced by these operators, the average earnings per man amounted to fifty cents per day.

Weaver and Rich Hill Placers

The Weaver and Rich Hill placers are in southern Yavapai County, a short distance northwest of Octave and from six to eight miles in air line east of Congress Junction.

This placer area is at the southern margin of the Weaver Mountains, which rise to over 5,000 feet above sea level, or to more than 2,000 feet above the adjacent desert plain to the south. Rich Hill stands at an elevation of 5,200 feet above sea level between the deeply eroded canyons of Antelope Creek on the west and Weaver Creek on the east. Since the higher portions of the Weaver Mountains receive at least eighteen inches of rainfall per year, these two south-flowing creeks generally have some water in their upper courses and are subject to torrential floods during the rainy seasons.

History and production: In the early sixties, certain Indians who came to trade at La Paz, near the Colorado River, reported the abundant occurrence of gold farther east. One of them was persuaded to guide a party, consisting of Capt. Pauline Weaver, Maj. A. H. Peeples, and others, to the locality. This party happened to camp at the base of Rich Hill, after their guide had deserted them on the desert north of Wickenburg. A Mexican of the party, while looking for their strayed animals, discovered loose gold nuggets on top of Rich Hill. This discovery led also to the finding of the placers on Weaver and Antelope creeks

This whole area soon became the scene of intense activity, and in five years, according to Hall,[48] produced about $500,000. The loose gold underneath the boulders and in the crevices of the rocks on Rich Hill was easily gathered, but more effort was required to work the bouldery gravels of Weaver and Antelope creeks by panning, rocking, and sluicing. As much as $40,000 is said to have been taken from a certain acre, and the production of the whole area, prior to 1883, was estimated by Hamilton[49] at $1,000,000. The town of Weaver, on Weaver Creek, flourished until about 1896, but is now marked only by crumbling ruins. Blake,[50] in 1899, stated that the score or so of men who were

[47] Oral communication.
[48] Personal communication.
[49] Hamilton, Patrick, Resources of Arizona. 1883.
[50] Blake, Wm. P., Report of the Territorial Geologist, in Report of the Governor of Arizona, 1899, p 60

working these placers from year to year were supposed to be recovering over $2,000 per month. The recorded production from 1905 to 1931, inclusive, amounted to about $83,975. For the year prior to June, 1933, the yield amounted to about $1,800.[51]

Geology: The Weaver Mountains, which are made up mainly of pre-Cambrian granitic and schistose rocks, overlain in places by Tertiary lavas, contain the Congress, Fool's Gulch, Octave, and numerous smaller gold-bearing veins. The placer ground covers an area of approximately eight by five miles. According to local people, the most productive portions, which were in the northern half of this area, included about ten acres on the top of Rich Hill; certain portions of the sides of the hill; the channels and benches of Weaver, Antelope, and other washes; and the gravel benches that lie between these washes.

Rich Hill, which rises steeply for about 2,000 feet above the plain, consists of rather intensely jointed granite. In places, it is traversed by thin, lenticular quartz veins which carry pyrite, galena, and gold. The top of this mountain is a hilly mesa, about ⅞ mile long by ⅜ mile wide, that evidently represents an erosional remnant of the elevated Weaver Mountain pediment. It includes several acres of broad, shallow basins and drainage channels whose granite floors are mantled with granite boulders and very thin, rusty, sandy soil. A few angular pebbles of quartz and of hematite are locally present, but no alluvial or foreign gravels occur anywhere within the area. The once abundant occurrence of placer gold within the shallow basins and drainage channels is proclaimed by the numerous old workings that scoured every square foot of their surface.

Along the washes and benches below Rich Hill the placer material consists of iron-stained gravel and sand, up to ten or more feet thick, together with abundant subangular boulders that are two to six feet in diameter (see Plate 4).

Character of the gold: According to Heikes,[52] the fineness of the Rich Hill and Weaver placer gold is 910. On Rich Hill, according to Blake,[53] one nugget worth $450, and three worth a total of $1,008, were found. Mr. C. B. Hosford[54] of Octave, stated that the largest nugget found on upper Weaver Creek was worth $396, and that two chunks of quartz contained $450. In the spring of 1931, a large nugget was brought into the office of the Arizona Bureau of Mines from the Weaver region. This nugget was described by Heineman[55] as follows: "The nugget is in general outline shaped somewhat like a human molar. It measures

---

[51] Written communication from Mr. Carl G. Barth, Jr., of Octave.

[52] Heikes, V. C., Dry Placers in Arizona: U S. Geol. Survey Mineral Resources for 1912, Part I, p. 259.

[53] Blake, Wm. P., work cited.

[54] Oral communication.

[55] Heineman, Robert E. S., an Arizona gold nugget of unusual size: American Mineralogist, Vol. 16, No. 6, June, 1931, pp. 267 to 269

approximately 53 mm. across the widest portion of the "roots," and 47 mm. from the bottom of the "root" to "the crown." Several fragments of slightly iron-stained quartz remain in the center of the mass. The total weight is 270.90 grams, and it may be calculated that the nugget consists of 252.38 grams of metal and 18.52 grams of quartz . . . worth $152.62 in gold and 22.71 grams of silver worth 21 cents at date of writing."

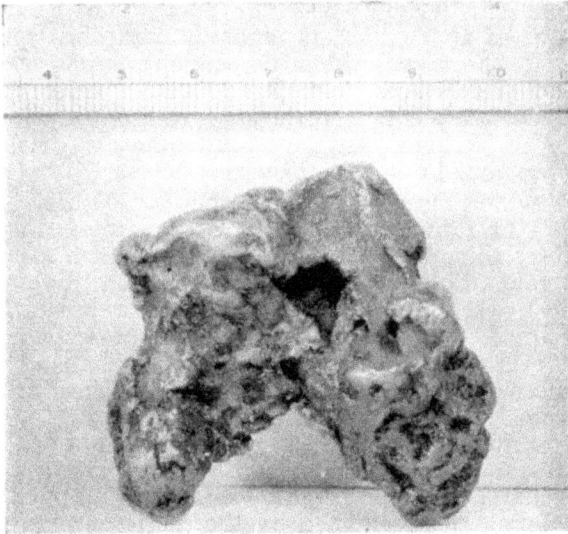

Photograph by Robert E Heineman

Plate 3. Nugget from Weaver placers. Natural size

During the 1932-33 season, a few nuggets ranging up to more than three ounces each in weight were obtained from Weaver Creek. Two nuggets, each weighing more than five ounces, were found on upper Antelope Creek.[56]

Away from the margin of the mountains, coarse gold becomes progressively more rare.

Origin: These placers were probably derived by the erosion of many small veins within the vicinity and cencentrated by local streams. Such large, angular boulders (see Plate 4) and such generally coarse gold could not have been transported far in ancient river channels.

Recent operations: Although this placer ground is privately owned, small-scale work, without royalty, has generally been allowed.

According to Mr. Carl G. Barth, Jr,[57] of Octave, approximately fifty men were carrying on sluicing and rocking in this field during the winter of 1932-1933, but their number decreased to

[56] Written communication from Mr Carl G. Barth, Jr
[57] Written communication.

eighteen with the advent of summer. Because the gravels are mostly coarse (see Plate 4) and have been repeatedly worked, the average daily earnings were not more than thirty cents per man.

Minor amounts of dry-washing have been carried on in the vicinity of Oro Fino Gulch, in the southern portion of the area.

Plate 4. Typical gravels of Weaver Creek placers.

COPPER BASIN PLACERS

The Copper Basin placers are north of Copper Basin Wash, between Skull Valley and the Sierra Prieta. They are accessible from the Santa Fe Railway at Skull Valley and Kirkland by a few miles of road. These placers, which have been intermittently worked in a small way for more than half a century, began to attract renewed attention in 1929.

Here, a plain slopes southwestward from an elevation of 5,500 feet at the base of the Sierra Prieta to 4,000 feet at the junction of Skull Valley and Copper Basin Washes. Most of this plain is floored with extensive deposits of gravel, sand, and clay, locally interbedded and mantled with volcanic tuffs and flows, but its easternmost one to three miles of width is a pediment that has been carved on granite. The whole area is dissected by many southwestward-trending gulches which are tributary to

Skull Valley Wash. Part of Copper Basin Wash carries a small flow of water throughout the year, but the other gulches are dry except for occasional short periods.

The bedrock of the placers generally consists of cemented gravels, but, in certain areas relatively far from the mountains, it is hard clay.

The gold-bearing gravels are made up largely of granitic sand together with various amounts of boulders and clay. Near the mountains, the boulders are relatively abundant and coarse but, in the western part of the area, they are mostly less than one foot in diameter and constitute a small percentage of the gravels. The clay content is erratically distributed, but tends to be relatively greater towards the western part of the area, except near Copper Basin Wash where sand predominates.

The gold-bearing gravels form a relatively thin mantle on the ridges, but range in thickness from three or four feet up to fifteen or more feet in the gulches. They contain some gold throughout their thickness but generally are richest in a thin streak at or near bedrock. Widely distributed tests indicate that much of the ground within this field contains from 50 to 83 cents in gold per cubic yard.

The gold, which is from 925 to 950 fine, occurs as particles that range in size from small specks up to nuggets several ounces in weight. In the western part of the field, nuggets worth more than 25 cents each are rare. Near the mountains, the gold fragments are characteristically wiry to angular and coarse.

Associated with the gold is abundant magnetitic black sand. In the upper portion of Copper Basin Wash, oxidized copper minerals are commonly present. Throughout the southwestern portion of the field, small particles of cinnabar (mercury sulphide) and natural amalgam, which were doubtless derived from the cinnabar veins of Copper Basin,[58] are apparent in the placer concentrates.

Origin: Erosion of gold-bearing veins of the Sierra Prieta, particularly in the pediment area, provided the gold of the Copper Basin placers. The increase in angularity and coarseness of the gold towards the mountains indicates a local derivation.

Recent production: The U. S. Mineral Resources credit the Copper Basin placers with a production of $1,023 during 1931. For the year prior to June, 1933, Mr. G. L. Lyda, of Kirkland, estimates the yield at about $31,000, of which $26,000 came from large-scale operations.

Recent large-scale operations: During the year 1932, three concerns carried on large-scale operations in the Copper Basin placers.

---

[58] These quicksilver deposits have been described by Carl Lausen and E. D. Gardner in Univ. of Ariz. Bureau of Mines Bul. 122, pp. 35-44. 1927.

Plate 5. Gold Star Company plant, Copper Basin district, in June, 1933

In the southwestern part of the field, the Forback & Easton and the Smith companies ran separate concentrating plants equipped with power shovels, trommels, screen, Diester-type tables and amalgamators which had capacities of 350 or more yards per eight hours. Water for these plants was pumped from shallow wells and reused as much as possible. The Forback & Easton plant closed down in the fall of 1932 and has been taken over by Mr R. Cassendyke. Its production is reported to have been from $12,000 to $15,000 worth of gold, most of which was in particles worth less than 25 cents. The Smith Company has been succeeded by the Gold Star Placer Company, also controlled by Mr. Cassendyke. Its plant, which was resuming operations in June, 1933, is illustrated in Plate 5.

During April and May, 1932, a lessee, equipped with a 1½-yard power shovel and a Girand barrel concentrator, recovered approximately $5,000 worth of gold from 6,000 yards of gravel on the Lyda ground in Mexican Gulch, about 2½ miles from Skull Valley.[59] Some $15 nuggets were found, but most of the gold ranged from $3 nuggets down to particles as small as a mustard seed.

In June, 1933, the Operators and Developers Company had installed in the northeastern part of Copper Basin a plant with a rated capacity of 500 yards per 24 hours. This plant is epuipped with a vibrating grizzly, a washing trommel, vibrating screens, sluice boxes, and Wilfley and Diester-type tables. Water is to be pumped from the Loma Prieta mine shaft which is about one mile farther south. The placer gravel, which is to be mined from an adjacent gulch, has been shown by partial tests to contain approximately 83 cents in gold per cubic yard.[60]

Recent small-scale operations: During the year prior to June, 1933, from fifty to sixty small-scale, individual operators produced approximately $6,000 worth of gold[61] from the Copper Basin placers. Most of this work was done with rockers (see Plate 11) and small sluices in Copper Basin Wash. According to Mr. A. S. Konselman,[62] of Prescott, the daily earnings per man ranged from 25 cents to $1.00 and averaged about fifty cents.

BIG BUG PLACERS

The Big Bug placer region is in south-central Yavapai County, in the general vicinity of Big Bug Creek, Mayer, Poland, McCabe, and Humboldt. This region includes a pediment at the northeastern foot of the Bradshaw Mountains and extends up certain gulches. Big Bug Creek, which empties into Agua Fria Creek, is perennial in approximately the upper half of its course.

---

[59] Oral communication from Mr. G. L. Lyda
[60] Oral communication from Mr. Homer Edwards.
[61] Estimate by Mr. G. L. Lyda.
[62] Oral communication.

Plate 6. Nugget from Big Bug placers, nearly actual size.

Photograph by Bate

History and production: Gold was discovered within the Big Bug region in the late sixties, but the greatest activity in placer mining there was during the eighties. Considerable sluicing, rocking, and panning have gone on, especially in upper Big Bug Creek as far down as Mayer, and in Chaparral and other gulches near McCabe. Dry-washing has been done to some extent in the drier portions of the region. In 1926, bullion having a fineness of 0.952 was recovered by sluicing operations of the Uncle Dudley Mining Company.[63] No estimates of the early production are available, but the 1910 to 1931 reported yield, as shown in the table opposite page 16, was $30,751.

Geology: The principal rocks of the Big Bug region are pre-Cambrian schists, smaller amounts of granite and granodiorite, abundant rhyolite dikes, and Tertiary basalt flows.

The placers occur in stream channels and on certain intervening mesas of a roughly triangular area that extends for about twenty miles east and northeast from the head of Big Bug Creek. The gold of the stream placers is generally coarse. One of the largest nuggets found in the Big Bug region contained about $500 worth of gold, and is illustrated in Plate 6. In the gravel mesa between Humboldt and Mayer, the gold, which is rather finely divided and associated with considerable clay, amounts to about thirty to forty cents per cubic yard.[64]

Certain quartz veins within the older rocks of the vicinity provided the gold for the stream placers, but the finely divided gold of the gravel mesas between Mayer and Humboldt may have undergone longer transportation.

Recent operations: According to Mr. F. W. Giroux,[65] of Mayer, the average weekly production of the Big Bug placers during recent favorable seasons has ranged from $200 to $300. During the fall, winter, and spring period of 1932-1933, sixty or more men were placer mining within the area, largely in the gravel benches and side gulches of Big Bug Creek, several miles northwest of Mayer. In this area, which has been rather intensively worked during the past, most of the mining is done by tunnels from which the gravels are packed to sluices, rockers, or small power-driven concentrating machines. Small-scale efforts are greatly handicapped by the large proportion of coarse boulders within the gravels.

Large-scale operations were attempted during 1932 by the Humphries Investment Company of Denver, with a large track-mounted power shovel, a Barber Green stacker, and sluices, but the enterprise was not successful.

In July, 1933, Pantle Brothers began large-scale operations on a 220-acre tract leased from Messrs. Shank and Savoy, west of

---

[63] U. S. Bureau of Mines, Min. Res., 1926, Pt. I, p. 667, 1928.

[64] Oral communication from Mr. Homer R. Wood, of Prescott.

[65] Oral communication.

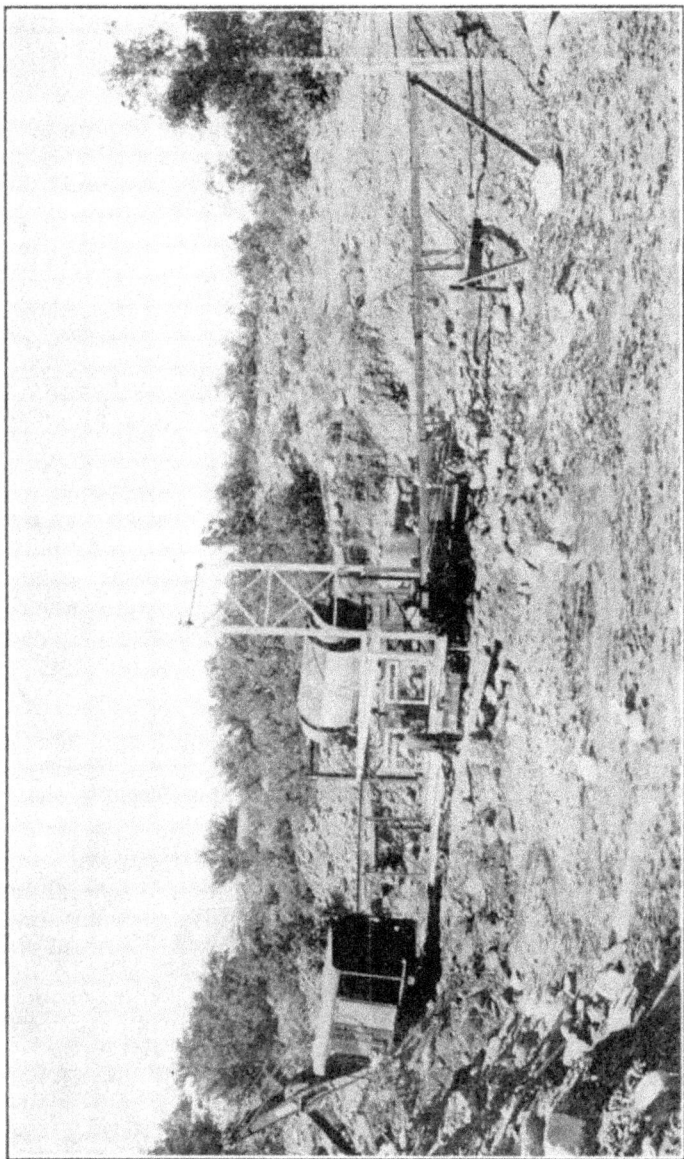

Plate 7. Ainley Bowl concentrator, Pantle Bros. lease, Big Bug Creek, during construction

Big Bug Creek and about three miles northwest of Mayer. When visited in August, 1933, this concern was mining old placer and mill tailings in a gulch near Big Bug Creek, where the ground is bouldery to sandy, with but little clay, and rests upon a bedrock of cemented gravels. The gold occurs as rather irregularly distributed, flat to round and ragged particles which range up to about fifty cents each in value.

The Pantle Brothers' cencentrating plant was equipped with four rubber-riffled Ainley centrifugal bowls (see Plate 7). It was fed with a one-yard power shovel, had a capacity of one cubic yard per minute, and required about 300 gallons of water per minute. Ample water for this plant was obtained at bedrock. Production during the first forty days of run amounted to about 45 ounces of gold.[66] Four men are employed.

HASSAYAMPA PLACERS

Placer gold occurs along practically the whole course of Hassayampa Creek in Yavapai County. The Creek rises in the Bradshaw Mountains at an elevation of approximately 7,000 feet above sea level, a few miles south of Prescott, and crosses the Yavapai-Maricopa County line two miles north of Wickenburg at an elevation of about 2,000 feet. Due to its large drainage area, this creek carries torrential floods in the rainy season, and abundant subsurface water during the dry months.

History: According to local reports, the greatest period of activity in the Hassayampa placers was from 1885 to 1890. The failure, in 1890, of the Walnut Grove Dam prevented the large-scale operations that had been planned for a tract downstream from Wagoner. Small-scale, individual sluicing and rocking have been carried on every year, but the total production therefrom is unknown. The U. S. Mineral Resources credit the Hassayampa placers with a yield of $1,336 in 1926, $1,208 in 1928, $485 in 1929, and $630 in 1931.

Geology: The principal rocks of the lower Hassayampa region of Yavapai County are pre-Cambrian granite and schist, mantled in many places by Tertiary gravels and lavas. In the upper or Bradshaw Mountain region, are pre-Cambrian schists and granite, intruded by smaller masses of diorite, granodiorite, and rhyolite porphyry. Pre-Cambrian to Tertiary quartz veins within the schists and granites provided the gold that erosion has concentrated in the placer deposits. The gold found along the upper reaches of the creek was generally coarse, but downstream it was progressively finer.

Recent operations: During the 1932-1933 season, more than fifty men were working the Hassayampa placers of Yavapai County. Most of this activity was confined to the side gulches. In general, the average daily returns amounted to about fifty cents per man.

---

[66] Oral communication from Pantle Bros

MINNEHAHA PLACERS

Placer gold occurs along Minnehaha Creek, about twenty-five miles in air line south of Prescott, below elevations of 5,000 feet above sea level. Lindgren[67] says: "Minnehaha Flat is a north-ward-trending, well timbered and watered basin on the head-waters of Minnehaha Creek, which discharges into Hassayampa River near Walnut Grove. Placer mining was carried on here in the eighties of the last century all the way up from the 'Old Log House' to the Button Mine, also in branches coming in from the east. The gold was worth about $17 an ounce and was extracted by arrastres, sluices, and dry-washers. The probable production was $100,000, according to Mr. M. A. McKay, an old-time resident of the district. The gold is believed to have been derived from the Fortuna lode near Lapham's place."

Placers on Oak Creek, below Fenton's ranch, have yielded a little gold during recent years.

GROOM CREEK PLACERS

The Groom Creek placers are in south-central Yavapai County along Groom Creek, from four to six miles south of Prescott. This creek heads in the Bradshaw Mountains west of Walker at an elevation of more than 5,000 feet above sea level and joins Hassayampa Creek at a point some five miles in air line farther southwest and some 1,900 feet lower.

These placers were discovered in the sixties, and were worked actively during the eighties. Their total production, according to former State Historian Hall,[68] probably has amounted to about $100,000.

Quartz veins contained within the local pre-Cambrian schist, which has been intruded by diorite, granodiorite, granite, and dikes of rhyolite porphyry, were the original source of the gold of these placers.

Recent operations: For several years past, only slight activity has been reported in the Groom Creek Placer field, and a very small amount of gold has been produced there.

PLACERITA PLACERS

The Placerita placers are in southwestern Yavapai County, about nine miles in air line south-southeast of Kirkland, in the vicinity of Placerita, French, and Cherry gulches. In 1899, Blake[69] stated that "The placers . . . at Placerita have long been known and worked, and are regarded as good-wages mines." Ac-

---

[67] Lindgren, Waldemar, Ore deposits of the Jerome and Bradshaw Mountains quadrangles, Arizona: U. S. Geol. Survey Bul. 782, p. 177. 1926.
[68] Oral communication.
[69] Blake, W. P., Report of the Territorial Geologist, in Report of the Governor of Arizona, 1899, p. 66.

cording to the late A. B. Colwell,[70] a dredging project was attempted several years ago on a rather small area of ground in French Gulch, about one mile below Zonia, but the dams failed.

No records or estimates of the production of this field are available.

The gulches of this vicinity have dissected a northeastward-sloping pediment whose general elevation is less than 5,000 feet above sea level. This pediment, which has been carved upon granite, diorite, and steeply dipping schist, is locally mantled by gravels and lavas. It contains many small gold-bearing quartz veins. Erosion of such veins probably furnished the gold of the placers.

Recent operations: When water was available during the 1932-1933 season, approximately 25 men were placer mining in the vicinity of the junction of French and Placerita gulches, chiefly with rockers and sluices. Their average daily earnings were about fifty cents per man. According to Mr. A. R. Evans,[71] of Kirkland, the production of this area for the year prior to June, 1933, amounted to approximately $2,000. This gold was fairly coarse, with many $5 and $10 nuggets and one $80 nugget. It is worth about $18 per ounce.

In the upper portion and side gulches of Placerita Creek, three or four men were operating long toms and dry-washers on shallow gravels. They each obtained from 25 to 50 cents worth of coarse gold daily.

Large-scale operations were started in June, 1933, on the Maude Lee claims, at the junction of Placerita and French gulches. The plant included a one-yard gasoline shovel, angle iron riffles, and a barrel amalgamator. Here, the gravels consist mainly of granitic sand with some medium coarse, flat schist boulders. A small flow of water occurs at bedrock.

MODEL PLACERS

The Model placers comprise a small area in the vicinity of Model Creek, on the western side of Peeples Valley. This locality is accessible by some two or three miles of road which branches westward from U. S. Highway 89 at a gate ¾ mile north of Peeples Valley store. These placers have been known for at least sixty years, since the discovery of the Model and other gold-bearing veins in this vicinity, but very little mining of them has been done during the present generation.

Here, a granite pediment extends, between elevations of 5,000 and 4,500 feet above sea level, from the eastern foot of the Weaver Mountains to Peeples Valley. This pediment has been dissected to shallow depths by several small eastward-trending streams which carry water only during part of the year. In certain places

---

[70] Oral communication.
[71] Oral communication.

between these gulches, it is concealed by thin granitic detritus and soil which supports a thick growth of brush and scrub oak.

Placer gold occurs in Model and other gulches for some distance upstream, but the principal gold-bearing gravels being worked in June, 1933, occurred in small, local basins or channels on the pediment for a width of about ¾ mile on each side of Model Creek, downstream from Pawley's place.

The placer gravels, which consist mainly of granitic sand with some clay and few boulders, are generally less than six feet thick. They contain a little gold throughout, but are richest in a six to twelve inch streak that rests upon granite or cemented granitic sand. Partial tests of this pay streak show about $1.00 worth of gold per cubic yard. The gold occurs as fairly rough particles that range up to about ½ ounce in weight and are reported to be about 850 fine. It was probably derived by erosion of gold-bearing quartz veins in the vicinity.

In June, 1933, approximately twelve men were engaged in small-scale placer mining operations in this field. After stripping off the overburden, the pay streak is carefully hand-shoveled and swept from the bedrock and hauled to Model Creek for concentration.

BLACK CANYON PLACERS

Placer gold occurs along Black Canyon, which drains the water of Turkey, Poland, Bumblebee, and several other creeks southward into the Agua Fria River. According to Lindgren,[72] "Placers have been worked at several places in Black Canyon, particularly below the Howard Copper Company's property. A few years ago a Portuguese is said to have taken out $20,000 near the old stone cabin, one mile below Howard. There are also small placer deposits near Turkey Creek station, and every year more or less dry washing is done by Mexicans in this locality."

The placer gravels in much of this field contain abundant coarse boulders. The gold particles are generally flat, fairly coarse, and worth about $18 per ounce. Black sand occurs abundantly in the gravels and adheres to the smaller gold particles.

Recent operations: During the cool portion of the 1932-1933 season, about 25 men, mostly transients, were engaged in small-scale placer mining in Black Canyon, chiefly between Arrastre Creek and Cleator, and to a small extent in American and Mexican gulches. Most of the concentrating was done with rockers and sluices, and only a small amount with dry-washers. The average daily returns were very small. Mr. W. J. Martin, store-keeper at Bumblebee, purchased approximately $80 worth of gold per month and estimates that an equal amount was marketed elsewhere. The largest nugget found during the past year came from American Gulch and was worth $14.38.[73]

---

[72] Work cited, p. 157.
[73] Oral communication from Mr. Martin.

On a bar some three miles south of Bumblebee, a plant equipped with a power shovel, screens, and tables was operated for a short time during the summer of 1932.

GRANITE CREEK PLACERS

Placer gold occurs along the upper branches and main course of Granite Creek, which rises a few miles south of, and flows northward through Prescott. These placers were discovered in the sixties, and were worked south of Prescott to a considerable extent during the eighties. New England Gulch, a branch of Granite Creek, about four miles south of the city, was very rich. According to Mr. Homer R. Wood,[74] of Prescott, some small nuggets have been found in digging excavations for buildings in that city. Lindgren states[75] that a little placer gold has been mined also at Del Rio, about 22 miles north of Prescott.

The U. S. Mineral Resources credit the Granite Creek district with a production of $390 worth of placer bullion in 1931.

EUREKA PLACERS

Gold placers occur in Burro Creek and other gulches of the Eureka district of western Yavapai County, about eighteen miles in air line northwest of Hillside. According to Mr. Homer R. Wood,[76] of Prescott, more than one hundred men were dry placer mining at the old Placers, near the Cowboy Mine, during the late fifties. The U. S. Mineral Resources record a placer production of $363 from the Eureka district in 1914, and a little in 1922.

HUMBUG PLACERS

Regarding gold placers in the Humbug district of southern Yavapai County, Lindgren[77] says: "The Humbug district, adjoining the Tiptop on the west, contains many gold-bearing veins, but most of its production evidently came from placers, now exhausted, in Swilling, Carpenter, and Rockwall gulches, which are small tributaries of Humbug Creek."

According to Mr. C. L. Orem, gold-bearing gravels occur for more than twenty miles along Humbug, French, and Cow creeks. These gravel bodies are generally less than a few hundred feet wide and range up to about twenty feet in maximum thickness. Their texture ranges from fine to very coarse. The gold tends to be flaky or floury in the upper gravels and rather coarse at or near bedrock. It is worth $18 per ounce in the local market.

Recent operations:[78] During the past two years, more than 200 men, mostly transients, have carried on small-scale placer opera-

---

[74] Oral communication.
[75] Work cited, p. 54.
[76] Oral communication.
[77] Work cited, p. 178.
[78] Written communication from Mr. C. L. Orem.

tions in this region. The average daily returns per man were less than fifty cents. In 1933, the rainfall was insufficient for sluicing but greatly hindered dry-washing, and only a few men were operating.

PIMA COUNTY

The principal gold placer districts of Pima County are Greaterville, Quijotoa, and Las Guijas. Other placers are known elsewhere in the county, but, except in the Old Baldy, Alder Canyon, and Papago districts, they have been of little or no economic importance.

GREATERVILLE PLACERS

The Greaterville district is in southeastern Pima County, at the eastern foot of the Santa Rita Mountains. The village of Greaterville, which is in the approximate center of the placer area at an elevation of 5,280 feet above sea level, is about 34 miles in air line southeast of Tucson and 8½ miles northwest of Sonoita, a station on the Nogales-Benson Branch of the Southern Pacific Railroad. This district is accessible by several short roads that branch west from the Tucson-Patagonia highway.

The Santa Rita Mountains, which attain, in Old Baldy Peak, 7½ miles southwest of the camp, an elevation of 9,432 feet above sea level, receive abundant rainfall and are well timbered. Although this rainfall varies somewhat from year to year, the average annual amount for elevations of 4,000 to 6,000 feet above sea level is over fourteen inches, and for elevations over 6,000 feet is from sixteen to more than twenty inches. About 75 per cent of the precipitation occurs in July, August, September, and October, and a large part of the other 25 per cent falls during the winter as snow. The eastward-sloping placer region is dissected by numerous steep-sided, nearly east-west gulchs which drain to Cienega Creek and are about one hundred feet deep near Greaterville. The only perennial stream of the district is situated about four miles south of the village. Sufficient water for domestic purposes, but not for much gravel-washing, is obtained from shallow wells in Empire, Ophir, Kentucky, and Big gulches

History: According to Raymond,[79] placer gold was discovered in the Greaterville district in 1874 by A. Smith. From 1875 to 1878, the placers were worked by 200 or more men.[80] The virgin gravels are said to have been so rich that each man recovered $10 or more daily by rocker with water packed in for four miles on burros and retailed at about three cents per gallon. After 1880, due to the richer gravels having been worked over, activity in the camp declined, and by 1886 had practically ceased.

---

[79] Raymond, R W , Statistics of mines and mining in the states and territories west of the Rocky Mountains for 1875, pp. 389-390. 1877.
[80] Hinton, R. J., Handbook of Arizona, p 213 1878.

According to Schrader and Hill,[81] sluicing was carried on in Kentucky Gulch for a few months during 1900. In 1902, considerable ground was owned and operated by the El Oro Mining Company. By 1905, the Santa Rita Water and Mining Company had begun operations on about 2,000 acres of patented ground. Their hydraulicking equipment included eight or ten miles of ditch and pipe line from a system of dams in Gardner and South canyons in the mountains. Profitable operations were conducted by them for a short time.

Further hydraulic operation[82] were tried by another company, at the junction of Kentucky and Boston gulches, with a 125-foot head of water brought through an eight mile pipe line from the first canyon south of Gardner Canyon. Considerable sluicing of the creek bed is reported to have shown, however, that the gravels in the overburden there were rather coarse and the returns too low to warrant further work.

Another company installed a one-ton steam shovel, screens, and a conical concentrating tank in Empire Gulch just below Enzenberg Canyon, but the pay dirt was not rich enough to warrant the removal of the sixteen or more feet of overburden.

Production: According to Raymond,[83] the yearly production of the Greaterville placers from 1874 to 1883 was estimated at $12,000. Burchard[84] places the 1884 output at $18,000. From 1902 to 1931, the production of the district reported by the U. S. Mineral Resources, as shown opposite page 16, totaled $42,756. The L. E. Jones Company of Greaterville, reports purchasing $67 worth of placer gold from the district during the last half of 1925, $182 during 1926, and $32 up to May 20 in 1927.

Geology: The accompanying map (Figure 6), after Hill[85] and Schrader shows the general geology and distribution of placer gravels in Greaterville vicinity. In the vicinity of the larger intrusives, there has been considerable local metamorphism that is marked by sericitization and silicification. Near Granite Mountain, the sedimentary beds are strongly impregnated with quartz, and sericite, together with some calcite, pyrite, and chalcopyrite. Here also are gold-bearing quartz veins that probably gave rise to the placers. East of the Cretaceous belt are eastward-thickening, imperfectly stratified, very angular gravels and sand that have been derived by erosion from the Santa Rita Mountains. This material commonly is cemented by clay or lime car-

---

[81] Schrader, F. C., Mineral deposits of the Santa Rita and Patagonia Mountains, Arizona: U. S Geol. Survey Bul. 582, p. 159. 1915.

[82] Schrader, F. C., work cited.

[83] Raymond, R. W., statistics of mines and mining in the states and territories west of the Rocky Mountains, p. 342. 1876.

[84] Burchard, H. C., Production of the precious metals in the United States, p. 46. 1884.

[85] Hill, J. M., notes on the placer deposits of Greaterville, Arizona: U. S. Geol. Survey Bul 430. p. 12. 1910.

bonate. It is dissected by many broad, deep-sided gulches, and contains the gold placers of the district.

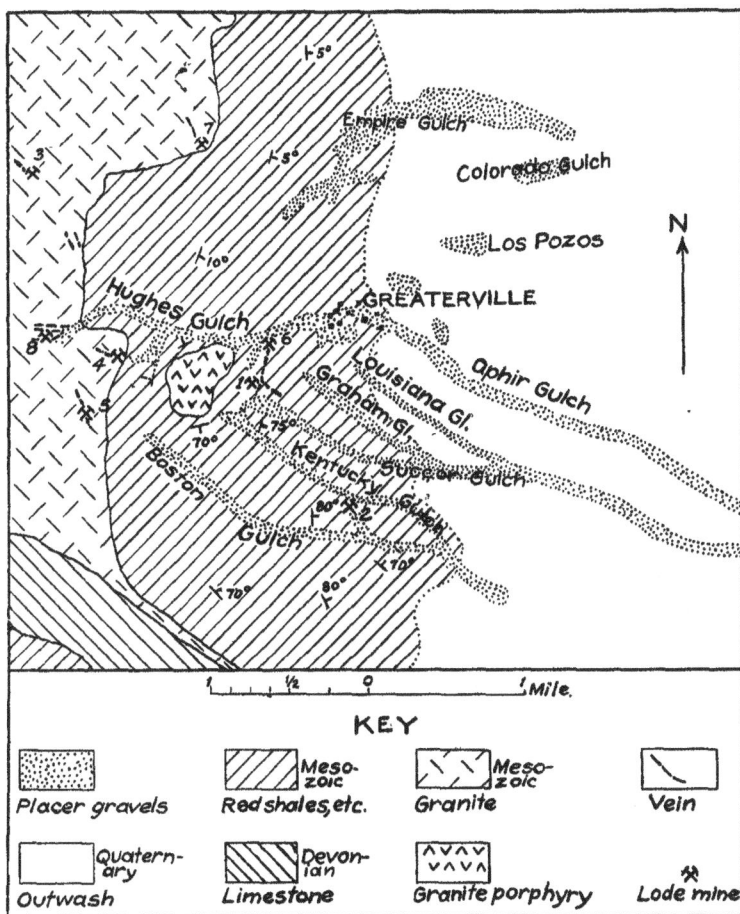

Figure 6. Geologic map of the Greaterville placer region, after Schrader and Hill, with certain alterations. Lode mines: 1, Fulton; 2, Harshaw; 3, Mountain King; 4, Quebec; 5, Royal Mt.; 6, St. Louis; 7, Wisconsin; 8, Yuba. Devonian area includes also other Paleozoic rocks.

Character and distribution of the gravels: Schrader[86] gives the following description of the gravels:

"They are irregularly distributed, chiefly in the bottoms of the present stream courses and gulches, where the principal diggings occur in shallow ground, and also upon the benches, slopes, and tops of the ridges, where some of them seem to represent deposits in old stream channels, examples of which occur just south of Greaterville thirty feet above the valley, on the crest of the

[86] Schrader, F. C., work cited, p. 161.

ridge to the southeast, and on the north side of Hughes Gulch below the mouth of Nigger Gulch fifteen feet above the bottom. They consist chiefly of a two-foot bed of angular gravel which rests uncomformably upon the bedrock of all the different older formations contained in the area, including the early Quaternary cement rock. They are covered by one foot to twenty feet or more of overburden composed of later Quaternary and recent gravels and wash. In places, as in Kentucky, Ophir, and Empire gulches, the upturned, irregularly eroded edges of the underlying sedimentary beds form natural riffles, behind which the gold has been concentrated.

"The gravels of the gold-bearing bed are generally small, the pebbles, as a rule, being less than an inch in size, though in many places cobbles four to eight inches in diameter occur. In a few places the gravels are crudely stratified and slightly cemented, generally by lime. They are sharply angular and but slightly water worn. The sand consists chiefly of angular fragments, and many of the particles of quartz and feldspar show well-preserved crystal faces. The coarse material consists chiefly of red and yellow sandstone, shales of various colors, arkose, a little dense white rhyolite, and granite porphyry. The gravels rest in most places on a red-brown clayey matrix which is handled without difficulty by hydraulic methods."

Character of the gold: "The gold, which is rather uniformly distributed throughout the bed, is mostly coarse. It ranges from flakes one-tenth of an inch in longest diameter, which was the size of most of the material recovered at the time of the visit in 1909, to nuggets worth a dollar or more. The gold of the early days was all coarse, nuggets ranging from $1 to $5 in value being common. Some nuggets brought into Tucson contained from $35 to $50 worth of gold, and the largest nugget reported from the camp weighed 37 ounces and had a value of about $630   The gold averaged about $17 to the ounce fine, and it was not difficult for a man to take out an ounce a day. The gold, like the containing gravels, is very angular, with many pointed projections, denoting that it is of local origin and has not traveled far. A little quartz adheres to some of it and seemingly also galena, both of which are reported to have been common in the large nuggets. The gold is mostly bright, but some of it is iron-stained and concentrates from panning contain considerable magnetic black sand."

According to L. E. Jones Company of Greaterville, a nugget worth $228 was found in 1924.

Productive gulches: Schrader[87] says:

"The productive gulches were Boston, Kentucky, Harshaw, Sucker, Graham, Louisiana, Hughes, Ophir below its junction with Hughes, the upper parts of Los Pozos and Colorado, Chispa

---

[87] Schrader, F. C , work cited, pp. 162-164.

on the road from Enzenberg camp to Greaterville, and Empire below its junction with Chispa.

"Boston Gulch: In Boston Gulch, which heads in the col south and west of Granite Mountain and trends a little south of east, gold was found in paying quantities from its head a point about half a mile south of its junction with Kentucky Gulch at the Kentucky camp. In the upper two miles of its course the gold was found in a channel five feet wide on bedrock, at two to four feet below the surface. Below Harshaw Gulch the gold was still confined in a ten-foot channel in the valley bottom, five to ten feet below the surface. Below the mouth of Kentucky Gulch the valley is wide, and for half a mile below this point the gold was distributed on bedrock at a depth of ten to sixteen feet for a width of approximately fifty feet.

"Harshaw Gulch: In Harshaw Gulch, a short, narrow tributary of Boston Gulch with steep bedrock sides, the pay streak, which in places was rich, was confined to the bottom of the gulch, about four feet wide.

"Kentucky Gulch· In Kentucky Gulch, which heads south-southeast of Granite Mountain and joins Boston Gulch at Kentucky camp, the gold occurs throughout its length on bedrock in a channel six to ten feet wide. At the upper end of the gulch the pay streak lay at the surface, but the covering gradually thickened to six feet at the mouth of the gulch.

"Sucker Gulch: In Sucker Gulch, which has three small heads southeast of Granite Mountain, the gravels were productive to a point a little below its junction with Ophir Gulch. From its head to the mouth of Graham Gulch the pay channel was six to nine feet wide and three to twelve feet below the surface. Between Graham and Louisiana gulches the pay channel averaged from twenty to fifty feet in width and the depth was from twelve feet at the former to 25 feet at the latter gulch. Below the mouth of Louisiana Gulch the gold was found distributed through the gravels on bedrock for a breadth of 100 feet. The overburden at the lower end was excessive, and therefore but little work was done.

"Graham Gulch· In the lower end of Graham Gulch, a short branch of Sucker Gulch heading southwest of the St. Louis mine, the pay gravel covered the entire bottom, about 100 feet in width, on bedrock at twelve feet below the surface. At the upper end of the gulch the pay streak was ten feet wide and was covered by only six inches of soil. Some gravels fifteen feet above the bottom of the gulch on the south side were also productive.

"Louisiana Gulch: At the head of Louisiana Gulch, which heads about a quarter of a mile south of Greaterville and joins Sucker Gulch a little more than a mile below, gold was found almost at the surface, but near the mouth of the gulch it lay at a depth of ten to twelve feet. The average width of the pay streak was about six feet.

"Hughes Gulch: In Hughes Gulch, which heads two miles west of Greaterville, just south of the Yuba mine, and extends north of Granite Mountain, a narrow channel, rarely over six feet wide from its head to its mouth, was found productive at two to six feet below the surface.

"Nigger and St. Louis gulches: Nigger and St. Louis gulches, small tributaries of Hughes Gulch, the first named lying to the west and the second to the east of Granite Mountain, contain small gold-bearing gravel channels.

"Ophir Gulch: Ophir Gulch, which heads northeast of the Yuba Mine, contains no placer deposits above its junction with Hughes Gulch. Below Greaterville, however, a channel 200 feet wide was found to contain gold as far down as the mouth of Sucker Gulch. The bedrock is rather deep here and little work has been done.

"Los Pozos Gulch: Los Pozos Gulch, which heads about a mile northeast of Greaterville, contains workable gravels in the upper 3,000 feet of its course.

"Colorado Gulch: In Colorado Gulch, a short branch of Empire Gulch, half a mile north of Los Pozos Gulch, some gold was found at shallow depths through a distance of 2,000 feet in the upper part of its course, nearly to its head.

"Chispa Gulch: In the lower three-quarters of a mile of Chispa Gulch, a small branch of Empire Gulch heading southwest of Enzenberg Gulch, a five- to ten-foot pay streak on bedrock at about ten feet below the surface yielded very high returns and was being worked at the time visited in 1909. In the lower portion of an east branch of Chispa Gulch gold was also being obtained from gravels three feet below the surface. At the head of the western fork of Chispa Gulch, which is about a mile in length, pay dirt lay at the surface, but at the mouth of the fork the gold was contained in a fifty-foot channel on bedrock with ten feet of overburden.

"Empire Gulch: In Empire Gulch placer gold was found only along a mile and a half of its course below the mouth of Chispa Gulch. The gold occurs in a bed two feet thick resting on conglomerate bedrock and is covered by sixteen feet of overburden. Near the mouth of Chispa Gulch the pay gravels were about 300 feet in width, but at the lower end of the pay belt they were distributed over a width of a thousand feet."

Origin of the placer gold: Since most of the productive gulches head in the Cretaceous sedimentary belt that surrounds Granite Mountain, the placers very probably were derived mainly by erosion of quartz veins of that vicinity. These veins have been prospected in the Yuba (Inghram), St. Louis, Quebec, and other lode mines, and found to contain more or less free gold. Particularly in the Yuba, some beautiful wire gold has been found. That the gold of the placers has not been transported far from its ultimate source is proclaimed by the angularity of its flakes and nuggets.

Recent operations: A few men carry on intermittent, small-scale placer mining in the Greaterville district by digging pits or shallow shafts to bedrock and gophering out the gold-bearing gravels. The pay dirt is concentrated in rockers, with water packed from wells, but the net returns are very low. Due to the presence of clay in much of the gravel, dry-washing is not very practicable here. Much of the ground has been reworked several times, but a large amount of gold still remains in these placers. Due to such factors as overburden, clayey matrix, and lack of abundant local water supply, this gold can be recovered profitably only on a large scale, by dredges or by certain adequate hydraulic methods, after ample water supply has been developed. Because of these facts, large-scale placering operations have been contemplated by the Gadsden Purchase, Inc., the Greaterville Dredge Gold Mining Company, and other concerns.

Gadsden Purchase, Inc., which is the successor of the Santa Rita Water and Mining Company, has control of several thousand acres of ground in Hughes, Colorado, Los Pozos, Hefty, Ophir, Succor, Louisiana, Kentucky, Boston, Harshaw, and Fish gulches. This ground, according to the late M. E. Young,[88] contains from forty to sixty cents gold per cubic yard. The company planned to bring water, from reservoirs in Cave, Gardner, and Sawmill canyons, through about 12½ miles of ditches, tunnels, and steel pipe lines. Hydraulic, drag-line, and dredge operations were contemplated.

During the 1932-1933 season, from ten to twenty men carried on small-scale placer mining in the Greaterville district. The average daily returns per man were less than fifty cents. Due to a shortage of water, activity fell off considerably during 1933. The Jones store in Greaterville purchased approximately $2,200 worth of gold during 1932 and $900 worth during the first eight months of 1933.

QUIJOTOA PLACERS

The Quijotoa gold placer district is in the vicinity of the Quijotoa Mountains of central Pima County, about seventy miles west-southwest of Tucson. According to Stephens,[89] the placers cover probably 100 square miles, and Heikes [90] states that they extend north and south for some distance on both sides of the Mexican boundary.

The Quijotoa Mountains, which rise to about 4,000 feet elevation above sea level, or approximately 1,500 feet above the surrounding plains, extend from Covered Wells on the north to South Mountain on the south, or to within about twenty miles

---

[88] Oral communication.
[89] Stephens, Bascom A , Quijotoa mining district guide book: Tucson Citizen Pt. & Pub. Co 1884.
[90] Heikes, V. C., Dry placers in Arizona: U. S. Geol. Survey Mineral Resources for 1912, Part I, p. 260.

of the Mexican line. This region has a very hot climate in summer, and there is no water supply except from wells and from earth or rock tanks. The mean annual rainfall in the placer area is probably about ten inches.

History: There is no record of how long these placers have been known, but, in 1774, according to Elliot's History of Arizona (1884), a Castilian priest named Lopez carried on extensive mining in an area about six miles north of the Quijotoa Mountains. It is said that Lopez utilized the docile Papagoes for his work, and that the Mexicans who continued mining there until 1849 washed the gravels with water brought by Papago squaws from tanks in the valleys. For many years after 1849, there was little activity in the placers; but, in the early eighties, a very lively boom in lode mining attracted thousands of men to the district, and caused four or five towns to spring up. As this boom subsided, many of the men turned to placering, and there has been a small amount of activity ever since.

In 1906, the Imperial Gold Mining Company was said to own most of the productive ground, and to be leasing to dry-washers.

In 1910, a Quenner pulverizer and a Stebbins dry concentrator are reported to have been installed by the Manhattan Company in the Horseshoe Basin area, but, due to the difference in conditions from those existing where these machines had been successful, the experiment failed.

Production: Considerable gold was recovered from the Quijotoa placers during the early days, but there is no record of the amount. In 1899, Blake[91] was informed that "The placer mines in the near vicinity of Quijotoa, worked by the Papagos in their crude way, are producing annually between $6,000 and $7,000 worth of gold." As shown in the table opposite page 16, the U. S. Mineral Resources record a production of $29,906 from the district between 1902 and 1913. Only a small amount per year has been recovered since 1912.

Geology: The Quijotoa Mountains, which are made up mainly of granite and lavas, contain numerous deposits of gold, some of which locally contain small, rich pockets. Erosion of the gold-bearing rocks furnished the material for the placers. Much of the placer ground is reported to average over eighty cents per yard, and Stephens[92] states that the red colored dirt averages $5 a ton. This last figure, however, is probably too high for the area as a whole. In general, the gold is coarse.

In Horseshoe Basin, which is a pediment area four or five miles long by a mile or so wide at the eastern foot of the range, south of Covered Wells, the gold occurs erratically distributed for several feet down from the surface. The bedrock here is ce-

[91] Blake, Wm. P., Report of the Territorial Geologist, in Report of the Governor of Arizona. 1899, p 64.
[92] Stephens, Bascom A, work cited.

mented gravel or caliche. On the western side of the range, no cemented gravels have been reported.

Recent operation: During the cool portion of the 1932-1933 season, approximately 200 men came to the Horseshoe Basin area to mine placer gold, but most of them remained only a short time. In June, 1933, only a few men were carrying on intermittent dry-washing there. The average daily returns per man were very low. All of the ground is privately owned.

Placer gold has been mined by Papago Indians from an area about three miles south of Pozo Blanco and one mile west of the foot of the Quijotoa Mountains. The best gravel, which was about five feet thick, occurred at depths of twelve or fifteen feet and rested upon caliche. Mr. Miles Carpenter states that prospecting below this caliche has revealed damp clayey gravel which is locally rich in coarse gold. In 1933, a group of twenty individuals held a tract of eight square miles in this area.

During the late portion of the summer of 1933, the planning of large-scale operations on the western side of the range was announced.

LAS GUIJAS OR ARIVACA PLACERS

Las Guijas or Arivaca placer district is in southern Pima County, in the vicinity of Las Guijas Mountains and Arivaca, about fifty miles south-southwest of Tucson.

Las Guijas Mountains, whose rounded summits attain an elevation of about 4,400 feet above sea level, or about 1,000 to 1,400 feet above the surrounding plains, extend for about eight miles northwest from Arivaca. Temperatures in the summer are high and the mean annual rainfall is probably about fourteen inches. The drainage of the district flows northwest to Altar Valley through Arivaca and Las Guijas creeks. Arivaca Creek, which occupies a large channel along the southwestern foot of the mountains, contains water in its upper reaches during all of the year, but Las Guijas Creek, along the northeastern foot, is much smaller and drier. The district depends for its water supply upon shallow wells along the creeks and upon the flow of Arivaca Creek itself.

History: According to Bryan,[93] placers were being worked in Las Guijas Creek by Mexicans and Americans in the sixties and seventies. The name "Guijas," is Spanish for "rubble" or "conglomerate." Irregular, small-scale operations have been carried on for the past fifty years. Pits or shallow shafts are sunk to bedrock, and the few inches of richer material is then gathered up and treated in crude, hand dry-washers during the dry seasons, or in rockers after rains. Between 1890 and 1900, according to local reports, as many as one hundred placer miners occasionally worked in the district.

---

[93] Bryan, Kirk, The Papago Country, Arizona. U S. Geol. Survey Water-Supply Paper 499, p. 379. 1925.

Several projects for large-scale operations have been contemplated, but, so far, none have been successfully carried out. In 1915, the New Venture Gold Placer Company planned to pump water from Arivaca Creek, three miles away, for a special agitating sluice. This company had control of 4,200 acres of land, and asserted that each acre carried about 4,800 cubic yards of gravel worth $1 per cubic yard.

Production: No records of the production of these placers are available, but the total amount was undoubtedly large. Most of this yield was prior to 1900, and placer activity in the district gradually died down to practically nothing by about 1915.

Geology: The mountains of this vicinity, which are made up of lava flows, Cretaceous sedimentary rocks, and granite, contain gold-bearing veins that were the original source of the placers. The placer gravels have accumulated both on the pediment slopes or "mesas" and in the stream beds. Although the earliest placering in the district was mainly on the northeast side of the mountains, along Las Guijas Creek, gold-bearing gravels extend practically around the range, and in the gulches about Arivaca. Duzrano, Pisquero, Yaqui, and Sangose are the most noted gulches.

The mesa gravels contain some gold scattered throughout their maximum thickness of fifteen to twenty feet, but, in both the mesa and stream gravels, the highest values are at bedrock, or at clay-cemented false bedrock. In the mesa gravels, the gold is more angular and unpolished than in the stream beds, and commonly contains attached particles of the original gangue minerals. In general, the gold is rather finely divided, but, according to local reports, many of the nuggets were worth from $5 to $15, and one nugget valued at $192 was found in 1893.

Recent operations: During the winter of 1932-1933, approximately 100 men attempted placer mining in the gulches near Arivaca, but most of them were transients who won very little gold and remained only a short while. A few of the more experienced and industrious ones averaged about $1.00 per day. The gold particles generally range in size from flour up to that of a pin head and occur mostly at bedrock. Rocking, sluicing, and dry-washing methods of recovery were used.

In November, 1933, according to Mr. George R. Fansett, six men were placering with dry and wet methods in San Luis Canyon, midway between Arivaca and Buenos Aires. Here, the placers occur within certain areas on the inter-arroyo benches of a dissected pediment of sedimentary and volcanic rocks. The gravels, which contain some large boulders and in spots, considerable clay, are generally from two to six or more feet thick. The gold occurs mainly as fairly coarse, angular fragments. Part of the area is on State land.

At the same time, very little small-scale work was done in the northern part of Las Guijas placer area. During the winter of 1931-1932, the rains had been sufficiently heavy to enable about fifty men to conduct intermittent rocking and sluicing there. Due to the fine-grained character of the gold, dry methods of recovery are but little used in this ground.

In August, 1933, large-scale operations were being started by Arivaca Placers on the pediment at the northern foot of Las Guijas Mountains. This concern was experimenting with a concentrating unit equipped with a scrubber, screens, a Lamley jig, a table, and an amalgamation plate. Its capacity is rated at 150 cubic yards per ten hours and its water consumption at about 500 gallons per hour. Water was being obtained from an eighty-foot well in Las Guijas Creek. Gravel for this plant was being stripped from the pediment and from the gulches. It contains considerable sand and some boulders and rests upon Cretaceous shales. Much black sand and a little cinnabar are present in the gravels.

OLD BALDY PLACERS

The Old Baldy placer district is in southeastern Pima county, at the northwestern base of the Santa Rita Mountains, in the vicinity of Madera Canyon, about thirty miles south-southeast of Tucson. Of these placers, Schrader[94] says: "The Madera Canyon alluvial cone, heading near the foot of the mountains at an elevation of about 4,500 feet, slopes northwestward toward Santa Cruz River and has a radial length of at least five miles. It is composed of gravels and sands discharged from the mouth of the canyon. These gravel deposits in places are probably over 100 feet in thickness and they all carry colors of gold. Toward the head of the cone an eighty-foot shaft was sunk in them without reaching their lower limit. Below the road forks, however, the deposits are deeply trenched by recent gulches from forty to fifty feet in depth, some of which cut through the deposits to the underlying bedrock granite, and here considerable gold placer mining was done with fair returns in the early days, mostly in the late eighties, water being brought from Madera Creek by ditch and flume."

Recent operations: During the 1932-1933 season, the only activity reported in the Old Baldy placers consisted of sampling on the 28 claims held by the Onekama Realty Company.

PAPAGO OR AGUAIJITO PLACERS

Some placers are situated in the Papago mining district of southern Pima County, along Ash Creek on the Sunshine-Sunrise group of claims and in Pescola Canyon, about thirty miles southwest of Tucson. According to Allen,[95] "The area covered

---

[94] Schrader, F. C., Mineral deposits of the Santa Rita and Patagonia Mountains, Arizona: U. S. Geol, Survey Bul. 582, p. 180. 1915.

[95] Allen, M. A., Arizona gold placers: Univ. of Ariz., Bureau of Mines Bul. 118, p. 12. 1922.

by the auriferous gravel is small, but Mexicans working in the
rainy seasons are said to make good wages by the use of rockers.
There is ample water in the creek for the use of rockers then,
and the remains of old diggings indicate that a considerable
amount of work has been done there in the past."

BABOQUIVARI PLACERS

The following notes upon a newly developed placer field at
the eastern foot of the Baboquivari Mountains, five or six miles
southeast of Baboquivari Peak, have been supplied by Mr. George
R. Fansett. The gold-bearing gravels occur in benches and bars
along a large eastward-trending wash. In November, 1933, the
Edna J. Gold Placer Mines, Inc., held a lease on 680 acres of
State land in this vicinity and were installing a concentrating
plant equipped with a ⅜-yard power shovel and Lamley concentrators. Water for the project was to be pumped from a shallow
well. According to Mr. A. B. Conard, of the Edna J. Gold Placer
Mines, Inc., the company was planning to work a bar that contained approximately 50,000 yards of gravels which averaged 65
cents per cubic yard. The gold was reported to be rather finely
divided. The gravels were from six to eleven feet thick, with
abundant boulders but relatively little clay.

ARMARGOSA PLACERS[96]

Gold placers occur along the upper course of Armargosa Arroyo, which heads in the Tinaja Hills of southern Pima County,
six miles west of Continental. For several years, a minor amount
of dry-washing has been done in the gravels of tributaries to this
arroyo. After heavy rains, a little gold is recovered from the
thin soil and hillside detritus of certain portions of sections 20,
21, 28, and 29, T. 18 S., R. 12 E. Typically, these gravels are fine
grained. Partial tests of them by Mr. Arthur Jacobs, of Tucson,
showed 0.5 per cent of lead, 0.5 per cent of zinc, 0.8 oz. of silver,
and eighty cents in gold per ton. The gold is associated with
abundant magnetitic sand.

All of the ground in this area is privately owned.

ALDER CANYON PLACERS

Placer gold occurs in Alder Canyon, on the northern slope of
the Santa Catalina Mountains, from near the National Forest
boundary to within a few miles from the San Pedro River. These
placers have been known and intermittently worked in a small
way for many years. The gold-bearing gravels are reported to
occur as dissected bars or benches along the stream and to some
extent on the spurs between tributary gulches. The gold is
coarse, flat, and ragged.

---

[96] Acknowledgement is due Mr. Arthur Jacobs, of Tucson, for information
upon this placer area

During the 1932-1933 season, a maximum of fifteen or twenty men carried on rocking, sluicing, and dry-washing operations in this field. Most of them were transients who remained only a short while and won but little gold. Mr. J. W. Lawson,[97] postmaster at Oracle, purchased approximately $45 worth, near 936 in fineness, during the year. At present (August, 1933), all of the ground is privately owned and no work upon it by outsiders is permitted.

MARICOPA COUNTY

The principal placers of Maricopa County are in the Vulture, San Domingo, and Hassayampa regions. The annual rainfall of these regions is only about 10.5 inches, and the summer temperature sometimes is 113°. Their water supply during the dry seasons is from intermittent Hassayampa Creek or from wells, but the abundant, sub-surface seep of the Hassayampa has never been known to fail.

VULTURE PLACERS

The Vulture placers are in northwestern Maricopa County, in the vicinity of the Vulture Mine, about fourteen miles by road southwest of Wickenburg. North of these placers, the extensively dissected Vulture Mountains rise to elevations of 3,500 or more feet above sea level, or nearly 2,000 feet above the desert plain that adjoins the region on the south.

History: According to Mr. A. P. Irvine,[98] who spent many years in this district, these placers were first worked about 1867. At times during the five or ten years following, as many as two hundred or more men were placering with dry-washers in the arroyos of the vicinity. Blocks of ground only fifty feet square were allowed each miner, but many men recovered from $25 to $50 per day each. By about 1880, the richest, readily obtainable gold had been harvested, but some dry-washing, principally by transient miners, has been done every year after torrential rains. Evidences of the early activity along the arroyos are still to be seen in the numerous old pits, piles of screenings overgrown with small brush, and decaying dry-washer machines. In the northern portion of the area, even some of the thin hillside gravels were scraped up and dry-washed.

Geology: The principal rocks of the Vulture region consist of pre-Cambrian schists, dikes, and irregular masses of granite, probable Mesozoic monzonitic dikes, and Tertiary andesitic and rhyolitic lava flows. Within this schist are the large, rich, gold-bearing quartz vein of the Vulture Mine and many smaller veins. Practically all of these smaller veins carry visible free gold, and even the most minute drainage channels leading down from them contain placer gold.

---

[97] Oral communication
[98] Oral communication.

The Vulture placer ground covers about three square miles in the pediment of Red Top Basin, northwest of the Vulture Mine, and continues down Vulture Wash for about two miles southeast of the Vulture Mine. The placer gravels, which are composed mainly of medium to fine, angular pebbles of schist and quartz, are generally less than ten feet thick, and rest upon schist bedrock. Considerable caliche cement which occurs in all but the thinnest gravels, has limited dry-washing operations to the narrow arroyos that are typical of this field.

Although some gold is distributed throughout these gravels, it is more abundant near bedrock. Several samples, taken from random localities at the time of the writer's visit, revealed abundant colors when panned. Even the old dry-washer tailings show fine colors upon panning, for those machines could recover only the coarser gold. The gold is mostly coarse and angular. During the early days, according to Mr. Irvine,[99] many $10 to $20 nuggets were found, and some worth $100 were reported.

The origin of the placer gold, in Red Top Basin at least, appears to have been the small quartz veins of that vicinity. In this connection, Carl Lausen has observed that the gold of these veins, like that of the adjacent placers, is coarser than obtains in the Vulture vein. It is possible, however, that the gold in the drainage below the Vulture Mine may have been derived in part from the Vulture vein.

Present operations: At present, the only placering done in this region is by transient miners. The U. S. Mineral Resources report a production of $134 worth of placer gold from the Vulture district in 1931.

SAN DOMINGO PLACERS

The San Domingo region of northern Maricopa County adjoins San Domingo Wash, an eastern tributary of Hassayampa Creek, about forty-five miles northwest of Phoenix. This sharply and intricately dissected portion of the western foothills of the Wickenburg Mountains is from about 2,300 to 3,300 feet above sea level. It is traversed by a few roads from Morristown, a station on the Santa Fe Railroad.

History and production: The gold placers of this region were discovered many years ago. The greatest activity in the region is reported to have been between 1870 and 1880, when the towns of Old San Domingo and New San Domingo were maintained by the placer miners. About 1875, Old Woman Gulch, which is a southern tributary of San Domingo Wash, was a large producer.

Several projects have been planned for hydraulicking certain portions of this area. Dams have been proposed to catch the torrential run-off of the rainy seasons or to divert the sub-surface water of Hassayampa Creek. In 1910, a Mr. Sanger built a dam

---

[99] Oral communication.

across San Domingo Wash and started sluicing, but the reservoir filled up with sand and gravel before operations had proceeded for one season. Dry-washing, or rocking where there was enough water, has been carried on in the region every year since its discovery, and has supplied a large proportion of the placer production of Maricopa County recorded opposite page 16.

Geology: The principal rocks of this region are pre-Cambrian granites, gneisses, and schists, Tertiary basalts, andesites, rhyolites, agglomerates, and sandstones, and various dikes. Quartz veins, probably of both pre-Cambrian and post-Cambrian age, have furnished the gold that erosion has concentrated in the placers.

The placers occupy a belt, six or seven miles long by an irregular width, along the drainage system of San Domingo Wash. They are not confined to the stream beds alone, but are found also on some of the gravelly mesas that separate the gulches.

The gold itself is said to be angular, fairly coarse, and of 925 to 965 fineness. Several prospectors of the region state that, although much of the gold found was in pieces worth about $1, nuggets valued at $30 were common in the early days, and several worth $10 to $15 were found in 1925. The gold is reported to lie mostly near bedrock in the upper reaches of the gulches, but somewhat distributed through the gravels of the lower country, and to be associated with considerable black sand. Although the areas worked by the early-day dry-washers were rather rich, most of the ground is of too low a grade for such treatment. According to T. L. Carter,[100] part of the Lotowana Mining Company property along Rogers Wash was tested by over 200 holes, and an area of 300 to 350 acres, 2½ miles wide and 1,000 feet long, was found to range from one to twenty feet to bedrock and to average from forty to eighty cents per cubic yard. Sanger Wash was sampled by Mr. A. P. Irvine, of Wickenburg, and found to average 43 cents per cubic yard.

Recent operations: The San Domingo placers are still being worked to a small extent by dry-washers, or by rockers when there is enough water. Plate 8 illustrates the type of mining being done recently.

During the winter of 1932-1933, a large number of men attempted small-scale operations in the San Domingo placers, but most of them met with disappointment and remained only a very short time.

HASSAYAMPA PLACERS OF MARICOPA COUNTY

Gold is sparingly present in the gravels and sands of the whole Hassayampa River in Maricopa County and is notably abundant for a few miles below the mouth of San Domingo Wash, which

---

[100] Carter, T. L., Gold Placers in Arizona: Eng. and Min. Journal, vol. 91, pp. 561-562. 1911.

is about seven miles southeast of Wickenburg. According to Mr. A. J. Kellis,[101] of Wickenburg, who sampled a portion of this ground several years ago, bedrock is from fifty to seventy feet at the mouth of San Domingo Wash.

Plate 8. Typical placer work on San Domingo Wash.

COCHISE COUNTY

The best known gold placers of Cochise County are in the Dos Cabezas and Teviston districts. Other placers, of less economic importance, are known in the Huachuca, Gleeson and Bisbee districts, and a silver-gold placer occurs at Pearce.

DOS CABEZAS PLACERS

The Dos Cabezas placers are situated in north-central Cochise County, in the vicinity of Dos Cabezas village, at elevations of 5,000 or more above sea level. Allen[102] states that these placers were discovered in 1901 by some Mexican prospectors, but, although this discovery induced considerable local excitement, only a small amount of gold was recovered. During 1906, according to Heikes,[103] water was plentiful in the district for several months, so that considerable placer ground was worked by several companies and individual Mexicans. Many of the latter made from $4 to $6 per day with simply a gold pan. Some gold has been recovered from the Dos Cabezas placers practically every year since their discovery. The most productive years, as recorded by the U. S. Mineral Resources, were 1906, with $1,939; 1911, with $115; and 1914, with $228.

[101] Oral communication
[102] Allen, M. A., work cited, p. 19.
[103] Heikes, V. C., U S. Geol. Survey Mineral Resources for 1906, p. 155.

Practically all the gulches in the vicinity contain gold-bearing gravels. These gravels are rather thin in the canyons a short distance north of the village, but much thicker toward the south and away from the mountains. In places, sufficient clay is contained in the placer material to handicap extraction. The gold particles tend to be flat, ragged, and fairly coarse.

The abundant gold-bearing quartz veins and stringers that occur in the Mesozoic and older rocks of the Dos Cabezas Mountains appear to have been the original source of the gold.

Recent operations: During the winter and spring season of 1932-1933, approximately 25 men were placer mining in the Dos Cabezas district. Most of this work was done with the aid of dry-washers, but, in places, some small sluices were used.

TEVISTON PLACERS

The Teviston placers are in north-central Cochise County at the northern foot of the Dos Cabezas Mountains. Of these placers, Heikes[104] says, "During the wet season, dry-placer ground in the Teviston district yields a small quantity of gold yearly. About 300 acres have been reported valuable to a depth of from three to ten feet, the latter being the greatest depth prospected. Bedrock is from fifty to seventy-five feet in depth. Most of the gold is coarse, and the ground by tests has yielded from three cents to $28 per cubic yard. The largest nugget found was valued at $375. Some cement or caliche has been found in prospecting the ground, but values have been found in the gravel beneath."

This placer area is a pediment floored with granite and dike rocks and more or less mantled with soil and gravels. Many of the gulches leading out of the range contain placer gold. The gravels on the pediment consist largely of coarse to fine-grained granitic sand together with varying amounts of clay and a considerable percentage of coarse, semi-rounded boulders. Part of the gold, particularly away from the base of the mountains, is fine, but most of it near the mountains and in the mountain gulches is coarse.

The streams of this area are dry during most of the year, but water is usually obtainable at bedrock, from shallow wells or old mine shafts.

Recent operations: During the 1932-1933 season, particularly after heavy rains, a few men worked the gulches of the Teviston district for coarse placer gold.

The Gold Gulch Mining Company installed a "dry-land dredge" in the pediment area and carried on short experimental operations. This machine, which has a rated capacity of fifty cubic yards of gravel per hour and a water consumption of 150 gallons per minute, is illustrated in Plate 9.

[104] Heikes, V. C., Dry placers in Arizona: U. S. Geol. Survey Mineral Resources for 1912, Part I, p. 259.

Plate 9.   Gold Gulch Mining Company operations,
Teviston district, in June, 1933.

Huachuca Placers

Placer gold occurs in Ash Canyon of the southeastern portion of the Huachuca Mountains, about three miles north of the international boundary and twelve miles by road southwest of Hereford.

These placers, which have been known for many years, attracted very little attention until about 1911 when they yielded a nugget that contained approximately $450 worth of gold. This nugget is now in the collection of Mr. L. C. Shattuck, at the Miners and Merchants Bank in Bisbee. Since 1911, more or less small-scale sluicing has been carried on in Ash Canyon whenever water was available, but the production has been small and seldom was rocorded. For 1919, a yield of about fifty ounces of gold, including one 8½ ounce nugget, was reported.[105]

Placer gravels occur along the canyon bottom for a length of about three miles, mainly between elevations of 6,500 and 5,000 feet above sea level. Water is abundant in the creek except during dry seasons, when it is obtainable from springs and very shallow wells. The bedrock is cemented bouldery gravels in the lower reaches of the canyon and granite in the upper. The gold-bearing gravels contain numerous subangular boulders, up to several feet in diameter, and more or less locally iron-stained sand, but practically no clay. Most of the gold occurs at or near bedrock, but some is irregularly distributed throughout the gravels, and the $450 nugget is reported to have been found at a depth of only six inches below the surface. The gold ranges from flakes to rounded nuggets that are generally less than ¼ inch in diameter. It has not undergone long transportation and doubtless was derived from near at hand. Its original source may have been the numerous veinlets of coarsely crystalline auriferous quartz that cut the granite farther upstream.

Recent operations: During the winter season of 1932-1933, when water was rather plentiful, approximately thirty men were carrying on small-scale sluicing operations in the Ash Canyon placers. The average daily earnings were not more than 75 cents per man. All of the ground is privately owned. In June, 1933, three separate concerns were hydraulicking on a small scale, with water pumped from springs or shallow wells.

Gleeson Placers

During the year previous to June, 1933, several ounces of placer gold were produced by dry-washing operations near Gleeson,[106] central Cochise County. All of this ground is privately owned.

---

[105] Heikes, V. C., U. S. Geol. Survey, Mineral Resources for 1919, Part I, p. 342.

[106] For an account of the geology and copper deposits of the Courtland-Gleeson (Turquoise) region, see Univ. of Arizona, Ariz. Bureau of Mines Bul. 123.

The principal activity has been within an area, approximately ½ mile long by ⅛ mile wide, that lies 1¾ miles east of the post office. Here, a relatively thin mantle of gravel and soil rests upon a gullied pediment of limestone. This gravel consists of a loosely consolidated aggregate of feldspathic sand together with abundant pebbles and boulders of limestone and porphyry. According to Mr. G. C. Bond,[107] some of the gold occurs erratically distributed through the soil and gravel, but most of it is at the base of the soil. The gold occurs as particles which range in size from small specks up to nuggets worth $7.00 each. According to Mr. Bond, tests made on 100 cubic yards of this gravel showed it to contain an average of 66 cents per cubic yard. The fineness of the gold is 825. Abundant black sand and pebbles of hematite, as well as smaller amounts of oxidized copper, native silver, galena, and oxidized lead minerals, are associated with it. This placer apparently owes its origin to the erosion of gold-bearing quartz veins in the immediate vicinity.

In the gulch west of the Copper Belle mine, four or five men dry-washed for coarse placer gold during the summer of 1932.

Gold Gulch Placer, Bisbee District

Of the Gold Gulch placer, which is situated about four miles southeast of Bisbee, Ransome[108] says: "Small quantities of placer gold have been obtained from the upper part of Gold Gulch. This gold has been derived from the Glance conglomerate, and concentrated in the sand and gravel of the present arroyo. It is not present in sufficient quantity to be of economic importance."

Recent operations: In June, 1933, from ten to fifteen men were placer mining in Gold Gulch. The average daily earnings per man were reported to be twenty cents.

Pearce Placer

Some interesting information about the placer at Pearce, central Cochise County, has been furnished by Mr. Lewis A. Smith.[109] In 1895 this placer, which lies at the eastern and western margins of Pearce Hill, furnished the first carload of ore from the district. Further shipments, made between 1917 and 1927, have brought the total production of this placer to $8,700. The material, which has been derived by weathering of the quartz veins of Pearce Hill, is made up largely of boulders from a few inches to over three feet in diameter. It had a maximum thickness of twenty-five feet at the eastern margin of the hill

---

[107] Oral communication.
[108] Ransome, F. L., Geology and ore deposits of the Bisbee quadrangle, Arizona: U. S. Geol. Survey Prof. Paper 21, p. 121. 1904.
[109] Oral communication.

and fifteen feet at the western margin. The eastern margin averaged about twelve ounces in silver and $1.25 in gold per ton, while the western averaged fifty-seven ounces in silver and $15 in gold. These values were contained in manganese-stained, sugary quartz, and were present mainly as cerargyrite, embolite, and free gold.

GREENLEE COUNTY

CLIFTON-MORENCI PLACERS

Gold placers were discovered in the Clifton-Morenci or Copper Mountain district during the seventies, but, not being sufficiently rich to attract men away from the copper industry, they have not been intensively worked. The reported placer gold yield of this district from 1907 to 1931, inclusive, was $10,050, which came entirely from intermittent, small-scale operations.

Lindgren says:[110] "The gravels of the Gila conglomerate, resting in front of the older rocks on lower San Francisco River and Eagle Creek, are sometimes gold bearing, although the metal usually occurs only as very fine flakes. The late Quaternary bench gravels along the San Francisco above Clifton contain gold in a somewhat more concentrated form, and at Oroville attempts have been made to work them by the hydraulic method, but the results were not encouraging. This gold is probably derived from a system of veins outcropping on lower Dorsey and Colorado gulches, a few miles north of Clifton on the west side of the San Francisco River."

According to Blake,[111] a large sum of money was expended on a pipe line for the hydraulicking project near Oroville, but the want of adequate fall and space for the tailings caused the abandonment of the enterprise.

Lindgren continues: "Another gold-bearing district is that of Gold Gulch, two or three miles west of Morenci . . . About twenty years ago, the gulch was worked for placer gold."

Recent operations: In June, 1933, approximately 100 men were placer mining in the Clifton-Morenci district. All of the ground is privately owned, but small-scale, individual operations without royalty have generally been allowed. Water is abundant in San Francisco River and generally present in Chase Creek.

About fifty of these men were operating on upper San Francisco River, upstream from the pump station north of Clifton. In that vicinity, ancient river gravels rest upon granite bedrock some fifty or sixty feet above the stream. According to Mr. C E. Roark,[112] of Clifton, the richest material occurs as relatively thin

---

[110] Lindgren, Waldemar, Clifton Folio: U. S. Geol. Survey, Folio 129, p 13. 1905.

[111] Blake, Wm. P., Report of the Territorial Geologist, in Report of the Governor of Arizona, 1899, p. 66.

[112] Oral communication.

streaks in certain favorable channels at or near bedrock. This material is mined and carried to the river for treatment in sluices and rockers. In some cases, a preliminary concentration is made in dry-washers. The gold particles are generally coarse and assay about 850 in fineness.

Approximately eighteen men were carrying on small-scale operations on lower San Francisco River, mainly as far downstream as the mouth of Eagle Creek. This section of the river maintains a curved course that is deeply intrenched between bluffs of hard Gila conglomerate. Within the arcs of many of these curves are ancient river gravels which rest upon the Gila conglomerate. These gravels contain considerable sand and a large proportion of spheroidal boulders which are generally less than one foot in diameter. Black sand and abundant pebbles of magnetite, hematite, and limonite are present. These gravels, which range up to 25 feet or more in thickness, carry some gold irregularly distributed throughout but are generally richest at or near the Gila conglomerate bedrock. The gold particles range in size from flour up to that of a small bean. These placer gravels have long been mined by means of underground workings and treated in sluices and rockers at the river. During early 1933, sluicing operations were conducted on the Smuggler claims, twelve miles by road from Clifton. The gravels were pulled by a dragline scraper across a grizzly and into a 45-foot sluice for which water was pumped from the river. This sluice was lined with burlap, which Mr. T. M. Spencer, operator of the sluice, states,[113] is very effective in catching the finer gold.

Approximately 25 men were rocking and sluicing on Chase Creek between the old Rock House and Clifton. They obtained pay-dirt partly from tributary gulches, but mostly from ancient, elevated gravels resting on the Gila conglomerate. The average daily earnings per man were generally less than fifty cents. The gold in the tributary gulches tends to be fine-grained but, in the elevated gravels, its nuggets range up to about ½ ounce in weight.

In June, 1933, five or six men were rocking and sluicing in Gold Gulch, where water was available. The gravels there are relatively thin.

PINAL COUNTY

CANADA DEL ORO OR OLD HAT PLACERS

The only known gold placers of importance in Pinal County are in the vicinity of Cañada del Oro, in the Old Hat district. These placers, which extend also into Pima County, lie at elevations of over 2,600 feet above sea level, near the northwestern base of the Santa Catalina Mountains, from four to ten miles south of Oracle postoffice and 16 to 29 miles north of Tucson.

---

[113] Oral communication.

The water supply of this placer region is chiefly from wells and from the intermittent flow of Cañada del Oro Creek. The mean annual rainfall at Oracle, which is 4,500 feet above sea level, is about 19.44 inches but on the Santa Catalina Mountains, which attain 9,150 above sea level at Mt. Lemmon, less than ten miles southeast of the placer area, much heavier summer rains and winter snows obtain. Cañada del Oro, therefore, sometimes carries torrential floods during the summer, and a steady, small flow from the melting snows in the spring.

History and production: These placers are said to have been discovered by the Spaniards, during the early days of Tucson. Numerous old pits, trenches, and tunnels indicate considerable early placer mining, and many thousand dollars worth of gold are reported to have been recovered. The production recorded from 1903 to 1924, inclusive, amounted to $11,351.

Geology: The Santa Catalina Mountains are made up principally of pre-Cambrian gneiss, schist, and granite, Paleozoic sediments, post-Carboniferous granite, granite phophyry, diabase, and diorite, and Tertiary sedimentary rocks and lavas. Gold-bearing quartz veins, such as occur in the vicinity of the Copeland, Kerr, Matas, and other prospects in the upper reaches of Cañada del Oro, were the probable source of the placer gold.

Based upon information from Capt. J. D. Burgess, Heikes[114] describes the placers occurring in T. 10 S, R. 14 E., Gila and Salt River Meridian as having "apparently been deposited at intervals by floods from the Santa Catalina Mountains so as to form a deposit of nearly equal value from surface to bedrock, there being no pronounced accumulation of heavy gold at bedrock except in the stream, Cañada del Oro Creek, which passes through the region. The bed of dry gravel is from six feet deep at the creek side to 475 feet at the summit, with an average thickness of about 150 feet. The deposit is in general a loose gravel, uncemented. There are, however, alternating strata of deep-red, clayey material. These strata are of nearly uniform thickness of three to four inches and probably were formerly surfaces existing between floods, each being covered by a later flow of gravel from rainfall-eroded veins farther up the mountain. Shafts sunk on the hillsides from 27 to fifty feet in depth show values from ten to 42 cents per cubic yard. The average is difficult to determine, as the gold is not equally distributed. All the gold is found in well-rounded nuggets ranging from a few cents to $5 in value. There is a tradition of a lump weighing sixteen pounds with probably forty per cent of quartz, whose discoverers were found murdered in their camp sixteen miles north of Tucson. The nugget had disappeared. In fineness the gold averages about 905. Generally the placer material is dug, screened, and hauled

---

[114] Heikes, V. C., Dry placers in Arizona: U. S. Geol. Survey, Mineral Resources for 1912, Part I, pp. 259-260.

to the creek, and there worked by rockers, or sluiced when there is enough water. Many dry-washers have been tried, but most of the gold lies in the red clayey seams which apparently acted as bedrock for each period of deposition. Pulverizing this adherent material gives good results with the common bellows type of 'dry washer.' A boiler and pump were once used to throw water against the creek bank, but the water at that time proved insufficient for extensive operations."

Recent operations: During the 1932-1933 season, approximately thirty men intermittently carried on small scale rocking and panning in the Cañada del Oro region, chiefly on the northern side of the creek. Although one $25 nugget and a few $5 nuggets were reported, the averaged daily returns per man were seldom more than fifty cents.

In 1933, Gold Channel Placers, Inc., of Akron, Ohio, had control of 35 claims in T. 10 S., R. 14 and 15 E. and were planning large-scale hydraulicking operations.

SANTA CRUZ COUNTY

The best-known placer districts of Santa Cruz County are Oro Blanco, Patagonia, Harshaw, Tyndall, Nogales, and Palmetto.

ORO BLANCO PLACERS

The Oro Blanco placer district is in the Oro Blanco Mountains of southwestern Santa Cruz County, in the vicinity of Ruby and Oro Blanco, about 25 miles west-northwest of Nogales and a few miles north of the Mexican boundary.

The Oro Blanco Mountains, which attain in Montana Peak, near Ruby, an elevation of 5,500 feet above sea level, or about 2,000 feet above the deepest gulches, receive approximately fifteen inches of rainfall per year. The local water supply comes mainly from reservoirs or from shallow wells.

History and production: According to Mr. J. S. Andrews,[115] of Tucson, former storekeeper at Ruby, these placers produced about $2,000 worth of gold per year from 1896 to 1904, but this activity died down after 1907. Of the activity in 1899, Blake[116] says: "Most of the placer mining is carried on in a desultory way, often with a small and wholly inadequate water supply, and in certain places with dry-washing machines worked by hand. The returns are small, but the miners manage to get their living, especially where they can get water." An attempt at sluicing was made in 1906 by Kelly Brothers, but their earth-

---

[115] Oral communication.
[116] Blake, Wm. P., Report of the Territorial Geologist, in Report of the Governor of Arizona, 1899, p. 71.

fill dam washed out and caused the enterprise to fail. In 1911, only two properties were productive, and there has been very little activity in the placers of the district since 1915.

Geology: The Oro Blanco Mountains consist principally of pre-Cambrian granite, Cretaceous sedimentary rocks, Tertiary lavas, and various minor intrusives. They contain numerous gold-bearing quartz veins and stringers which have formed placers in most of the gulches that issue from the mineralized areas. According to Blake,[117] "In almost every ravine or gulch, gold can be found by panning, and even on the hillsides and on the surface generally, especially where the soil is reddened by decomposed pyrite, gold can be obtained by dry washing." Alamo and neighboring gulches, south and southwest of Ruby, contained the richest gravels. Mr. Andrews states that the placer gold was not very coarse, but ranged from flour up to one nugget worth $8. The fineness of the gold bought by Mr. Andrews from Old Oro Blanco was about $10 per ounce, and, from Alamo Gulch, about $16. The average fineness from the whole district was only about $12 per ounce, and the whiteness of the material containing this relatively high content of alloyed silver suggested the Spanish name "Oro Blanco" (white gold) for the district.

Recent operations: In the Oro Blanco region, small-scale placer mining activity depends upon the water supply. During the 1932-1933 season, the scarcity of water permitted less than the usual amount of work.

During 1932, the Gold Bar Placer Company installed a small scrubber and barrel concentrator in California or Oro Blanco Viejo Gulch, near the mouth of Warsaw Creek and about 2½ miles north of the international boundary. Water for this plant was pumped from a small reservoir in the canyon, and the gravels were obtained from a small basin floored with Cretaceous sedimentary rocks. The one short run that was made presumably failed to recover the fine gold present.

PATAGONIA OR MOWRY PLACERS

The gold placers of the Patagonia district, Santa Cruz County, are on the eastern slopes of the Patagonia Mountains, about nine miles south of Patagonia and six miles north of the Mexican boundary at an elevation of 5,200 to 5,800 feet above sea level. Of these placers, Schrader[118] says: "Placer gold occurs in the Patagonia district in the Quaternary stream gravels in the pediment portion of Mowry Wash and its tributaries, being present on the main wash at the east border of Quajolote Flat about 1½ miles southwest of Mowry, on a south-side tributary gulch about

---

[117] Blake, Wm. P., work cited.
[118] Schrader, F. C., Mineral deposits of the Santa Rita and Patagonia mountains, Arizona: U. S. Geol. Survey Bul. 582, p. 348. 1915.

1¼ miles south-southwest of Mowry, and on two north-side parallel tributary gulches about 1½ miles southeast of Mowry.

"The production is small, as the deposits are worked only by Mexicans when in need of money. The average earnings are about 75 cents a day for each man. The placers at the Guajolote locality were being worked by dry-washing at the time of visit (1909). The deposits at this place seem to be about five feet thick. The known production in 1909 was two ounces of gold. In 1906, when, after the closing of the Mowry Mine, many unemployed men were in the country, the production was about $200."

Recent operations: During the summer of 1933, approximately five men were carrying on rocking operations on Guajolote Wash, downstream from the old Mowry smelter. As water was rather scarce, the average daily returns per man were less than fifty cents. One two-ounce nugget and several smaller nuggets were found, but most of the gold occurs as angular particles, less than 0.1 inch in diameter, associated with abundant black sand.

HARSHAW PLACERS

According to Schrader,[119] "The only placers known in the Harshaw district occur about two miles southwest of Patagonia, between Sonoita Creek on the northwest and Alum Canyon on the southwest. Here the Quaternary gravels underlying the mesa-like area, which is about a mile square, contain placer gold and are workable under favorable conditions. They are said to contain also native lead. They were worked by A. J. Stockton and other pioneers by jigging in the early days."

TYNDALL PLACERS

Schrader[120] says: "Placer gold occurs in the Tyndall district, and some was produced in the early days 2¼ miles southwest of Salero and one mile south of Mount Allen, at the southwest base of Grosvenor Hills, on each side of the township line, in the S. W. ¼ Sec. 35 and adjoining ground, in the open basin-headed canyon which is tributary to Ash Canyon."

NOGALES PLACERS

According to Schrader,[121] "Gold placer deposits occur in the northeastern part of the Nogales district on Guebabi Canyon, which drains into Santa Cruz River from the northeast at a point about six miles north of Nogales. The canyon extends southwestward through a large area which is commonly known as the Guebabi district but which, except along the canyon, is

---

[119] Schrader, F. C., work cited, p. 279.
[120] Schrader, F. C., work cited, p. 220.
[121] Schrader, F. C., work cited, p. 355.

barren of ore deposits . . . . Along the course of the stream, gold placers of considerable extent are reported to occur in the Quaternary gravels . . . . The placers produced considerable gold in the early days, and were being worked to a moderate extent in 1909."

PALMETTO PLACERS

In 1927, the Patagonia Placer Mines Company had control of 320 acres in the Palmetto district and planned to recover placer gold from the Quaternary gravels at a point about 2½ miles northwest of the Three R. Mine or six miles by road southwest of Patagonia. This company installed a sluice and a drag-line excavator adjacent to the bed of the main arroyo, but, after a month of intermittent work, abandoned the project.

GILA COUNTY

Placer gold occurs in the Dripping Spring, Barbarossa, Globe-Miami, Pinto Creek, Payson, Mazatzal, and Spring Creek regions of Gila County. These placers have been mined in a small way since the seventies, but most of their production was during the early days.

DRIPPING SPRING PLACERS

The Dripping Spring placers occupy a small area northwest of Cowboy Gulch, on the southwestern side of Dripping Spring Wash. This area is a few miles west of the Globe-Winkelman highway and 24 miles from Globe.

These placers have been known and worked in a small way for half a century. Their yield, according to Mr. Cal Bywater, owner of the ground, amounted to about $3,000 in 1927, but was considerably less during most years.

The gold-bearing gravels are from twenty to eighty feet thick and rest upon hard Gila conglomerate. They contain few boulders more than one foot in diameter and are weakly cemented with red clay. Not much black sand is present. In the Southern portion of the area, the richest material is at or near the base of these gravels, but near the northern margin, it is erratically distributed. The ground that has been worked averaged about fifty cents per cubic yard.

About ten per cent of the gold is finer than 100 mesh, but the remainder occurs as fairly well-rounded nuggets which range up to approximately ½ ounces in weight. This gold is about 845 fine. It was probably derived by the erosion of small gold-bearing quartz veins in the adjacent mountains.

Recent operation:[122] During the winter of 1931-1932, more than 25 lessees earned their living in this area. Eight lessees worked there during the winter of 1932-1933. The gravel was

---

[122] Oral communication from Mr. Bywater.

mined from shafts, tunnels, and underground stopes and concentrated with water pumped from the United Vanadium Corporation's well. Good results were obtained by a small plant equipped with a shaker screen, amalgamation plates, and sluice boxes.

BARBAROSSA PLACER

Ransome,[123] in 1923, stated that "At a locality known as the Barbarossa mine, 2¼ miles southeast of Troy, free gold to the value of a few thousand dollars, probably from $2,000 to $3,000, has been obtained by dry washing the soil and loose detritus on the Troy quartzite. One nugget is reported to have weighted about 22 ounces. The nuggets showed little rounding and presumably were supplied by the disintegration of some small vein close at hand."

This locality is at an elevation of 4,000 feet on the southwestern slope of the Dripping Spring Range, opposite the Dripping Spring placers. During the season of 1932-1933, a few lessees were working the ground.

GLOBE-MIAMI PLACERS

Pinal Creek: Placer mining has been done along Pinal Creek, upstream from the town of Globe. According to Carl Lausen,[124] the nuggets were generally worth from a few cents up to 25 cents each, and a few $5 ones were found.

Gap and Catsclaw Flat area: Placer gold occurs within an area about 4,000 feet long by 1,500 wide east of Sixshooter Creek, some six miles southeast of Globe. During the early sixties of the past century, according to local reports, placer mining was carried on in this area with water packed from wells several miles away. After a short time, activity ceased until 1932 since when a few individual dry-washers have operated. All of this ground is privately owned.

Here, a pediment, carved mainly upon diorite and partly mantled with gravels, gives way northeastward to Gila conglomerate. The placer gravels occur mainly upon the diorite and to a minor extent upon the Gila conglomerate. They are generally fine to sandy, with minor clay and few large boulders. Their gold occurs as irregularly distributed, medium to fine, angular grains associated with abundant black sand. Partial tests by Mr. F. H. Chadwick, owner of part of the ground, indicate that the gravels in portions of the area run about 25 cents per ton.[125]

Richmond Basin: Probably some placer gold was recovered from the small gulches that drain westward from the Apache Mountains through Richmond Basin, north of Globe. This basin is noted for its rich placers of horn and native silver.

---

[123] Ransome, F. L., U. S. Geol. Survey, Ray Folio (No. 217). 1923.
[124] Oral communication.
[125] Oral communication.

Lost Gulch and Pinto Creek: For many years placer gold has been recovered from Lost and Gold gulches and from Pinto Creek, west-northwest of Globe. Blake[126] states that "Placer deposits of considerable extent and value have been worked for years in Lost Gulch, Globe district. These deposits appear to have been supplied by the disintegration and erosion of a multitude of small veins traversing the region." The U. S. Mineral Resources occasionally report a production of placer gold from Lost Gulch. The gold occurs in a rather spotty fashion both within the creek channel and the adjoining dissected benches. It ranges from fine to fairly coarse, and the largest nugget found here is reported to have been worth $43. During 1932 and 1933, approximately eight men spent part of their time placer mining with dry-washers, rockers, and sluices in Lost Gulch. Their average daily returns were low.

Golden Eagle: Certain gulch gravels east of the Golden Eagle vein, a short distance north of Miami, contain finely divided gold. In June, 1933, according to Mr. J. W. Strode,[127] about twelve men were dry-washing these gravels and making from fifty to sixty cents per day, each.

PAYSON PLACERS[128]

Considerable rich float from the gold-bearing veins of the Payson district, northern Gila County, was picked up during the seventies and eighties.

Although the quartz veins of the district show free gold at the surface, placers are not common. One short tributary of the East Verde River drains the region in which most of the gold veins occur; yet the prospectors of the district state that no placer gold has been found in it. Placers, however, have been worked in a small way for a number of years below Ox Bow Hill, but only during the rainy season when water is available. These gravels are worked occasionally and yield but low returns. On the slopes of Ox Bow Hill immediately below the outcrop of the vein, Mr. Boozer panned about an ounce of gold. Some of this gold consisted of rather coarse, flat nuggets up to a quarter of an inch in length. These nuggets are of a deeper color than the vein gold, and probably contain little or no silver. Mr. Boozer stated that any pan of the dirt from the slope will show a few colors.

GRAHAM COUNTY

GILA RIVER PLACERS

Placer gold occurs in eastern Graham County, along the Gila River, chiefly upstream from the mouth of Bonita Creek. The

---

[126] Work cited, p. 66.
[127] Oral communication.
[128] Lausen, Carl, and Wilson, Eldred D., Gold and Copper Deposits near Payson, Arizona: Univ. of Arizona, Bureau of Mines Bul. 120 1925.

western part of this area is accessible by seven miles of unimproved road which branches northward from the Safford-Duncan highway at a point about fourteen miles from Safford. These placers have been known and occasionally worked for about twenty years but have produced very little.

Here, the curved course of the Gila River is deeply entrenched between terraced bluffs of Gila conglomerate. Within the arcs of certain curves, these terraces are mantled with ancient river gravels which carry placer gold. The gravels, in general, contain a large proportion of boulders which range from several inches up to three feet in diameter. Ferruginous chert pebbles are fairly common, and black sand is very abundant. The gold which is flaky to wiry in form, ranges in size from that of flour up to wiry particles ¼ inch long. Partial tests indicate that the ground contains from fifteen to fifty cents per cubic yard.

Recent operations: During the past few years, interest in these placers has revived, and a small production from them has been reported.

At the Neel property, which is on the north side of the river between Bonita and Spring creeks, test-runs were made with a washing plant for which water was pumped from the river. In June, 1933, this ground was held by the Rio Gila Gold Mining Company.

Sampling has recently been conducted farther upstream, on the Smith-Boyle, Hammond-Serna, and Colvin properties.

MOHAVE COUNTY
GOLD BASIN PLACERS

Situation: The Gold Basin Placers of northwestern Mohave County are in T. 28 and 29 N., R. 17 and 18 W., about nine miles south of the Colorado River. Their central portion is accessible by about nine miles of unimproved road that branches northward from the Kingman-Chloride-Pierce Ferry highway at the northern end of Red Lake playa, 56 miles from Kingman.

History: The first known discovery of placer gold within this area was made in May, 1932, by Mr. W. E. Dunlop. In August of that year, approximately 100 men were testing the field with dry-washers. Most of them left during the winter rainy season, but about forty were there in June, 1933. As most of these people were transients who took part of their gold elsewhere, any approximate estimate of the production is difficult to reach.

Topography and geology: Gold Basin is floored largely by a detrital fan that slopes eastward from the White Hills to Hualapai Wash. This fan is approximately six miles long from west to east by five miles in maximum width. Its vegetation consists principally of small desert shrubs and abundant Yucca or Joshua trees. Water for all purposes is hauled chiefly from Patterson Well, several miles away.

The gold-bearing gravels occur principally in the arroyos and gulches, between elevations of 3,300 and 2,900 feet above sea level. They consist mainly of medium-grained, angular schist and gneiss fragments together with a minor amount of finely divided quartz. A small proportion of boulders, generally less than two feet in diameter, is present. The placer gravels are mostly from one to three feet thick and rest upon a bedrock of firmly cemented gravels. Their gold occurs partly as flour gold and partly as angular fragments that range from five cents to $3.50 in value. Some of the gold is attached to black schist particles. Black sand is rather abundant.

The tests that have been made of this ground show that the gold is rather erratically distributed. Certain pockety channels contain thin streaks that run more than $1.00 per cubic yard, but most of the arroyo banks probably contain less than $1.00 per cubic yard. The cemented gravels of the bedrock are reported to carry a little gold, but no test of them has been made.

Origin: The White Hills, which are made up of granitic, schistose, and volcanic rocks, contain many argentiferous and auriferous quartz veins.[129] Erosion of such veins doubtless gave rise to the Gold Basin placers. The occurrence of most of the gold as angular fragments, some of which are attached to black schist particles indicates some such nearby source.

Operations: Prior to June, 1933, the operations in this field consisted of small-scale, individual dry-washing. A few more experienced, industrious workers each made $1.00 or more per day, but most of the operators averaged less than that amount. Dry-washing here is interrupted during the rainy seasons.

During the summer of 1933, a large-scale dry-treatment plant was installed by Mr. S. C. Searles in Sec. 29, T. 29 N., R. 18 W. This plant, which is equipped with a grizzley, a trommel, screens, and a battery of twelve dry-washers, has a rated capacity of twenty cubic yards of gravel per hour.

All of the known placer ground in the Gold Basin field is controlled by a few individuals and companies, part of whom lease claims to the small-scale operators. The results of this practice have not always been satisfactory, particularly in certain outlying sections where the gold is too finely divided for economic recovery.

KING TUT PLACERS

Situation: The King Tut placers of northwestern Mohave County are in T. 29 and 30 N., R. 17 W., about eight miles from the Colorado River. They are accessible from Kingman, via Chloride and the Pierce Ferry Highway, by 72 miles of improved road.

---

129 Schrader, F. C., Mineral deposits of the Cerbat Range, Black Mountains, and Grand Wash Cliffs, Mohave County, Arizona: U. S. Geol. Survey Bul. 397, pp. 127-135. 1909.

Plate 10. Motor-driven dry-washer, Searles group, Gold Basin district, in June, 1933.

History: So far as is known, the first discovery of placer gold within this area was made in February, 1931, by Mr. W. E. Dunlop. According to Mr. Charles Duncan,[130] the only gold production has been incidental to sampling and has amounted to about $700. All of this land is privately owned, chiefly by the Duncan ranch and by the Santa Fe Railway.

Topography and geology: Here, a gravel-floored plain, from 3,000 to 4,000 feet above sea level, rises southwestward between Grapevine Wash and the base of a low northward-trending ridge locally called the Lost Basin Range. Near these mountains, the plain is a pediment floored with schist and granite. Its vegetation consists principally of small desert shrubs and abundant Yucca or Joshua trees. Water for all purposes is hauled from Patterson Well, five miles distant.

The richer, gold-bearing gravels, as known in June, 1933, occur within an area some eight miles long by an undetermined width and are confined mainly to the arroyo-bottoms. They consist predominantly of slabby schist pebbles, with few boulders more than ten inches in diameter, intermingled with abundant silt and sand. These deposits are generally less than two feet thick and rest upon caliche-cemented gravels. Their gold occurs partly as fine material and partly as flat, ragged nuggets that are known to range up to 1/16 ounce in weight. Black sand is abundantly associated with it. Northeastward, the gold particles and the gravels become progressively finer grained.

Tests of part of the field showed average values of 69 cents per cubic yard.[131] According to Mr. Duncan, the underlying cemented gravels are also gold-bearing, but no comprehensive test of them has been made.

Origin: The King Tut placers probably originated from erosion of a group of gold-bearing quartz veins in the Lost Basin Range. The raggedness of the gold nuggets, many of which carry attached quartz, indicates a local derivation.

Operations: Operations in this field prior to June, 1933, consisted of sampling and experimentation. Most of the testing was done with dry-washers. A few small wet machines were tried, but the water for them was found to be too costly.

On the Robeson and Joy lease, in sec. 14, T. 30 N., R. 17 E., a Cottrell dry concentrator with a capacity of 25 tons of gravel per hour was being installed.

CHEMEHUEVIS PLACERS

The Chemehuevis placers of southwestern Mohave County are in the foothills of the Chemehuevis or Mohave Mountains, about eighteen miles southeast of Topock. This area is part of the

---

[130] Oral communication.
[131] See Ariz. Mining Jour., vol. 17, no. 5, p. 10. July 30, 1933.

Gold Wing mining district. Its climate is very dry throughout the year and rather hot during the summer. These placers have been worked intermittently by small-scale dry methods for many years. Probably the most activity has been in the Mexican or Spanish diggings, in the vicinity of the Red Hills, at the southwestern foot of the range. In general, the gravels are angular and free from large boulders. Where deep, they are cemented with lime carbonate. The gold is fairly coarse.

Recent operations: During the winter of 1932-1933, a maximum of thirty men were working at one time in the placers of the Chemehuevis Mountains, but most of them left with the advent of the hot weather. According to Mr. J. H. Jones, of Topock, their gold production during that period amounted to about $1,200.

A little activity was reported in Dutch and Printer's gulches, on the northeastern side of the range.

LEWIS PLACERS

The Lewis placer is on the patented property of the old Bi-Metal gold mine, three miles southwest of Kingman and ½ mile northeast of McConnico.

Here, a granite area about 300 feet in diameter, has been considerably mineralized with slightly auriferous pyrite. Regarding the Bi-Metal deposit, Schrader[132] says: "The free gold to which the deposits owe their value seems to have been derived from a considerable thickness of overlying mineralized rock. As this overlying rock became disintegrated and was removed by erosion, the fine gold liberated from it gradually worked into the underlying rocks in which it is now found. Below or outside of the oxidized zone of mechanical concentration probably only very low-grade ore occurs. In some small gullies or lines of drainage within or at the border of the area, where further concentration by flowing water has taken place, several tablespoonfuls of mostly coarse gold, of which some of the largest nuggets contained about half a dollar each in gold value, are reported to have been panned."

During the winter of 1932-1933, Mr. Al Lewis mined and sluiced the gravels from a small draw in this area. According to Mr. E. Ross Householder,[133] of Kingman, this material ranged in value from $1 to $5 per cubic yard and yielded about $900 in gold that was worth $20.21 per ounce.

WRIGHT CREEK PLACERS

Small gold placers occur in the upper reaches and tributaries of Wright Creek, in the northeastern portion of the Cottonwood

---

[132] Schrader, F. C., Mineral deposits of the Cerbat Range, Black Mountains, and Grand Wash Cliffs, Mohave County, Arizona: U. S. Geol. Survey Bul. 397, p. 137. 1909.
[133] Written communication.

Cliffs. According to Mr. E. Ross Householder,[184] of Kingman, intermittent, small-scale operations have been carried on here for the past decade, but the total production has been small.

LOOKOUT PLACERS

The Lookout Placers are in the Maynard mining district, near the northern end of the Hualapai Mountains, about six miles southeast of Kingman. Here, certain areas of shallow gulch and hillside gravels contain rough, wiry placer gold. Mr. E. Ross Householder,[185] of Kingman, states that one dry-washer in this area obtained about $150 worth of gold during the 1932-1933 season.

SILVER CREEK PLACERS

Some minor gold placers occur in the valley of Silver Creek, about six miles by road downstream from U. S. Highway 66 and five miles northwest of Oatman.

Here, an irregular pediment of volcanic rocks is overlain by a mantle of gravels which, in certain areas, contain a little placer gold.

During the winter of 1932-1933, the Gold Gulch Gravel Company attempted to work this ground with a large centrifugal bowl machine for which water was piped several miles. A short run, however, sufficed to determine that the gold present was insufficient to make the project profitable.

A short distance farther southeast, a little small-scale placer mining, chiefly sluicing in connection with assessment work, has been carried on. According to Mr. B. White, one of the operators, the gravels there are very firmly cemented with caliche and contain about 100 pounds of black sand per cubic yard. This gold is about 730 in fineness.

COLORADO RIVER PLACERS

The sands and gravels of the Colorado River, downstream from the mouth of the Grand Canyon, contain finely divided gold which several dredging and sluicing operations have attempted to recover. One of these enterprises is mentioned by Heikes[186] as follows: "The large dredge built in 1909 on Colorado River, near the Arizona side, opposite El Dorado Canyon, Nevada, was of the suction type . . . . It was built to work the sand bars and failed on first test to extract the fine gold. It was subsequently carried from its moorings by high water and wrecked during the spring of 1910."

River-bar placers: Minor amounts of coarse gold have been

---

[184] Written communication.
[185] Written communication.
[186] Heikes, V. C., U S. Geol. Survey Mineral Resources for 1910, Part I, p. 235.

recovered by small-scale operations in certain elevated bars that have been formed largely by tributary canyons.

At Willow Beach, which is 65 miles from Kingman on the Boulder Dam highway, one of these ancient bars contains the Sandy Harris placer. This bar covers an area of about 250 square feet near the outer bow of a curve in the Colorado River and rests upon an irregular surface of gneissic granite some 150 feet above the stream. It is made up of an unassorted aggregate of boulders, gravel, and sand. The boulders, which range up to more than six feet in diameter, are but slightly rounded and could not have been transported far. Likewise, the coarseness of the gold indicates a local derivation. This placer material was probably eroded from gold-bearing rocks in the vicinity and washed, by way of tributary gulches, to the river where it accumulated in the outer portion of the nearest curve. Subsequent downcutting of the river has left this bar elevated in its present position. Some 35 years ago, Mr. Harris worked this placer by tunneling on bedrock. In 1920, an unsuccessful attempt was made to sluice the gravels with water pumped from the river. The ground is now held by Mr. Al Jagerson, for whom a lessee has taken out about ten ounces of gold during the past year. Black sand is abundant in this placer.

Some medium coarse placer gold has been recovered from a bench near the Colorado River about 2½ miles north of Pyramid Rock.

COCONINO, APACHE, AND NAVAJO COUNTIES

The Triassic Chinle formation[137] of the Painted Desert in Coconino, Apache, and Navajo counties, northeastern Arizona, deserves mention as a low-grade gold placer that is of spectacular interest from a geological rather than a known practical point of view. This formation, which was known as the Shinarump prior to 1917, consists largely of mauve to variegated clays. It underlies the major portion of northeastern Arizona north of the Little Colorado River and outcrops as shown on the Arizona Bureau of Mines geological map of Arizona.

An account of the gold in the Chinle clays has been given by Lawson.[138] According to Lawson, these clays, when examined microscopically, appear to be composed almost wholly of a colloidal substance with a very small admixture of fine silt and some concretions of lime carbonate and iron oxide. He gives the fol-

---

[137] For a description of the Chinle formation, see Gregory, H E., U. S. Geol. Survey Prof. Paper 93, pp. 42-50. 1917.

[138] Lawson, A. C., The Gold in the Shinarump at Paria: Economic Geology, vol. 8, pp. 434-448. 1913.

lowing chemical analysis of the gray clay from Lee's Ferry, Coconino County:

SiO_2 .... 53.45
Al_2O_3 .... 18.56
Fe_2O_3 .... 7.89
CaO 1.87
MgO 1.66
H_2O at 105°C.. 6.77
Ignition loss above 105° C.. 7.26

97.46

When immersed in water, the clays swell enormously, break down rapidly, and run like milk. The mixture is in such a fine state of division that it passes freely through filter paper. When dried, the clay breaks down to an extremely loose, soft dust, but, due to the slight rainfall of this region, it has generally been disintegrated only to a depth of from one to two feet. Lawson found that the Chinle clays averaged five cents in gold per cubic yard at Paria, Utah, and states that they "appear to be similarly auriferous at Lee's Ferry . . . ; and it is probable from the extreme uniformity in the physical characteristics of the formation wherever it has been observed that it is similarly auriferous throughout its extent." The gold is in a very finely divided condition.

In June, 1933, a grab sample of disintegrated gray Chinle clay was taken by the writer from the base of a small knoll about five miles east of Cameron. A lump sample of the undisintegrated clay was taken from the Spencer property near Lee's Ferry, some 68 miles farther north. Assays by Mr. W. A. Sloan, of the U. S. Bureau of Mines, showed the Cameron sample to contain nine cents in gold per ton and 0.007 per cent of mercury and the Lee's Ferry sample four cents in gold and 0.041 per cent of mercury. It must be emphasized, however, that such near-surface samples may contain local concentrations of these metals and not be truly representative of the whole formation. As Lawson[189] says: "The value of the ground is very problematical. If a method of successful hydraulicking and recovering the gold be developed it will only be after a long period of experimentation, at large expense, at a few favored localities, where a vast yardage of the clays is free from overburden, and where abundant water may be had cheaply." Within the past thirty years, hundreds of placer claims have been staked out on the Chinle formation, but, so far as known, they have not produced any gold.

Recent operations: Since 1930, several concerns have sampled portions of this ground and tried out various extraction methods. In June, 1933, a company, headed by Mr. C. H. Spencer, was experimenting with a small hydraulic plant ⅛ mile north of the Colorado River, below Lee's Ferry.

---
[189] Work cited.

Plate 11. Rocker in operation, Copper Basin district.

PART II

SMALL SCALE GOLD PLACERING

By George R. Fansett

Mining Engineer, Arizona Bureau of Mines

INTRODUCTION

Gold, more than any other substance, is cherished by human beings. They are the only creatures who worship gold; it is the yardstick against which human beings measure all accomplishments. The story of gold is the story of mankind from the caveman era onward. Gold is the one commodity to which no sales problem is attached. Gold mining is the one industry which flourishes during a business depression.

The conditions existing on no two gold deposits are identical and the problems faced by any miner change from day to day, even on the same property. Such conditions require that the miner possess unusual ingenuity, resourcefulness, and open-mindedness. As all successful placer miners have utilized well known facts and established scientific principles in their placer mining operations, it stands to reason that, everything else being equal, the miner whose operations are based upon and conform to fundamental laws will make the best possible profit. For that reason, it seems to be advisable to stress some of them in this section of the bulletin.

FACTS ABOUT GOLD

Indentification of Placer Gold

Placer gold can usually be identified by noting certain of its characteristics, as follows:

1. Color.

Pure gold is brass-yellow in color, but as recovered from placers, it is usually alloyed with more or less silver and sometimes with copper. Silver tends to lighten the color without changing other characteristics, and a high percentage of silver makes gold silver-white with a slight yellowish tint.

2. Specific gravity.[1]

The specific gravity of pure gold is about 19.3. In other words, a given volume of pure gold is about nineteen times as heavy as

---

[1] The specific gravity of a substance is equal to its weight divided by the weight of an equal volume of distilled water at 4° centigrade (39° Fahrenheit).

an equal volume of water, about one and one-half times as heavy as mercury, more than twice as heavy as copper, about two and one-half times as heavy as iron, nearly seven times as heavy as quartz, and more than eight times as heavy as ordinary dry sand. Placer gold, almost always alloyed with silver, copper or other metals, has a specific gravity of from 15 to 19, depending on its fineness. It weighs from nine hundred (900) to eleven hundred and eighty seven (1187) pounds per cubic foot.

3. Malleability and ductility.

When gold is hammered on an anvil, it flattens out without cracking or breaking. A knife blade, needle, or similar tool cuts or indents gold in much the same manner as it does metallic lead. Iron pyrites or copper pyrites (fool's gold) and various other minerals that are often confused with gold are brittle and break easily when hammered. They can be reduced to a dark colored powder. Mica is much softer than gold, does not break easily when hammered, may be crushed with some difficulty to powder, and cracks readily.

4. Solubility.

Gold can not be dissolved in either nitric, hydrochloric (muriatic), or sulfuric (oil of vitriol) acid alone. It is soluble, however, in aqua regia which is a mixture of about one volume of concentrated (strong) nitric mixed with about two volumes of concentrated (strong) hydrochloric (muriatic) acid. Aqua regia solutions[2] of gold turn purple when stannous chloride is added to them. Ferrous sulfate, when added to such solutions, throws down a brown precipitate.

PHYSICAL PROPERTIES OF GOLD

Gold's high specific gravity (great weight per unit of volume) and its amalgamating characteristic are the two properties of gold that are utilized as the basis for all present-day placer mining methods.

Gravity concentration.

Due to its high specific gravity, gold, when suspended in water or air, settles faster than the other lighter minerals with which it is associated in the placer dirt. This makes it possible to concentrate, separate, and recover the gold from these lighter, worthless minerals. This property is made use of wherever the gold is won by panning, rocking, sluicing, dry-washing, or by any other gravity concentration method. Flaky, porous, or flour gold tends, however, to float in moving water, and, in dry-washing, it is apt to blow away with the tailings. Gold tends to float off in water if oil, grease, or clay is present.

---

[2] Detailed instructions for making these tests are given in the Arizona Bureau of Mines Bul. 128, Field Tests for Common Metals.

Amalgamation.

Amalgamation is the process of uniting mercury (quicksilver) with another metal. The amalgamation process that is used in gold placer mining is based upon the fact that, when clean, bright gold is brought into contact with clean, bright mercury, especially by a rubbing or grinding action, the mercury sticks to and coats the gold, forming an alloy. When particles of mercury-coated gold come in contact with each other, they become loosely cemented or soldered together; the resulting mass or paste is gold amalgam. Mercury will also amalgamate with copper and silver, but it will not amalgamate with quartz or granite.

Interfering factors.

If the mercury is dark or tarnished, the gold no matter how bright and clean it is, will not be caught or unite with the mercury. Neither will the union take place if the gold is dirty or rusty even if the mercury is bright and clean. Both the gold and the mercury must be bright and clean to amalgamate or unite. Grinding the mixture in cyanide solution (very poisonous) brightens the gold, cleans the mercury, and overcomes this trouble. Furthermore, if the mercury is broken up into many small globules or pellets (commonly called flour mercury), the union of the gold and mercury will not take place. When these globules of mercury are coated with a film, the mercury is then referred to as being "sick." Sick mercury will not amalgamate or catch the gold nor will these sickened mercury globules reunite or run together.

Mercury is tarnished by exposure to the air, especially after being in use for a time. It is also tarnished and coated by uniting with various substances among which are the oxides, sulfides, sulfates, and arsenides of the base metals (lead, iron, copper, antimony, etc.) and by certain minerals among which are talc and clay. Mercury is broken up and coated by various minerals that make up some "black sands" as well as by oil and grease. This is why the insides of pans, rockers, and other equipment used in gold placer mining, especially when amalgamation is employed, **must be free from oil and grease.** Agitating and washing the concentrates with lye or soda ash usually remedies the trouble that results from the presence of oil or grease, while cleaning the mercury by agitating it with weak nitric or sulfuric acid or cyanide solution usually causes the globules to reunite. The squeezing of sick mercury through chamois or canvas will not overcome this condition and it may have to be retorted in order to purify it. It can be seen, therefore, that sometimes real difficulties are encountered when the amalgamation process is used in gold placering. This is especially true when it is realized that the remedy that will overcome the trouble in one case may prove useless in another. In other words, each mercury flouring and sickening problem is a case unto itself and considerable testing may be required to find the correct remedy. The separation of gold from gold amalgam is described on pages 115-119 of this bulletin.

Size of Gold Particles

Gold, as found in placers, varies much in size, ranging from nuggets that weigh several ounces and even pounds to specks or colors that are commonly known as fine or flour gold. Some specks of flour gold are so minute that it takes as many as two thousand colors to weight enough to be worth one cent. The following classification of gold on the basis of size is from Young's[8] "Elements of Mining":

"Coarse gold—that which remains on a 10-mesh screen.

Medium gold—that which remains on a 20-mesh screen and passes a 10-mesh screen (average 2,200 colors to one ounce).

Fine gold—that which passes a 20-mesh screen and remains on a 40-mesh screen (averages 12,000 colors to one ounce).

Very fine gold—that which passes a 40-mesh screen (average 40,000 colors to one ounce)."

Flour gold—not defined, but presumably smaller than "very fine gold."

Purington quotes examples of finely divided gold as follows:

"170 colors to one cent (314,500 to one ounce).

280 colors to one cent (436,900 to one ounce).

500 colors to one cent (885,000 to one ounce)."

The grains of gold in a placer deposit are often much smaller than the grains of associated minerals.

The word "nugget" should be applied only to a piece of water-worn, native gold larger than a grain of wheat.

SEEKING PLACER GOLD

Since moving water has been the most potent factor in the development and formation of most placer deposits, the usual practice is to seek placer gold by panning along the water courses, namely, along stream beds, bars, gulches, and arroyos. Even though a placer of worth-while proportions and values may be situated far above any water course, nevertheless the showings which have been washed down from it to that water course, if followed up and traced out, may lead the prospector to the deposit. Black sand and the heavy minerals that accompany placer gold serve to some extent as a guide; the spots where heavy concentrations of them occur deserve special attention and testing, but it should be remembered that black sand and other heavy minerals are so excessively common that they can be panned from almost any soil. Large quantities of them mean, therefore,

---

[8] Young, George J., Elements of Mining, 3rd ed., McGraw-Hill Book Company, Inc., New York, 1932, p. 426.

merely that conditions were especially favorable to concentration. Gold is by no means always found where large quantities of black sand occur. All areas that look as though a slowing down or slackening of the water current has at some time taken place are worthy of testing since, in such areas, the rate of flow of the stream may have been so small as to cause it to drop the gold that it was transporting.

Gold, being heavier than most of the material with which it is associated, tends to settle and to sink to the bedrock. Bedrock and the dirt for a few feet above it should, therefore, be explored and tested with special care. Depressions in the bedrock may hold rich pockets of gold while bedrock that is fissured and shattered, acting as riffles, may hold good gold values.

A great many men, in the past few years, have been attempting to make a living in Arizona from the reworking of old placer ground in the old, richer fields. In such ground, the most easily found gold has been won, but, frequently, if the ground has not been worked over many times, the bottom has not been carefully searched, and the painstaking cleaning of crevices and potholes may yield lucrative results, especially where the bottom is soft or fissured. Evidences of such rich pockets are partly cemented rounded gravels and sand apparently forming a part of the bedrock. Any such suspicious part of the bedrock should be picked into for possible overlooked bonanzas. Such potholes and crevices may extend for several feet down into the bedrock and will almost invariably contain rich gravel if placer gold has been carried down that water-course.

Worn, rounded, smooth placer gold has traveled far from its mother lode, while sharp, angular placer gold or nuggets or grains that contain quartz or other brittle gangue minerals are comparatively close to their source. When seeking for the source of placer gold, the gold itself as well as "float" is followed up until no more is found. By trenching or sinking at that point, the mother lode may be opened up, providing it is still there— has not been removed by the processes that produced the float and placers.

PLACER EQUIPMENT AND METHODS

Gravity concentration.

As already stated, the great weight per unit volume of gold as compared to that of the other minerals with which it is associated in placer dirt is the factor that makes it possible to concentrate, separate, and recover placer gold. This property of gold is utilized as the basis of all gravity concentrating methods whether by the pan, batea, spoon, rocker (cradle), sluice, long-tom, dry-washer, or centrifugal machine.

Pan, miner's.

The gold pan (pan or miner's pan), Figure 7, is made from stiff sheet metal. Iron or steel is usually used, but aluminum and copper pans are available. Enamelware pans do not rust,

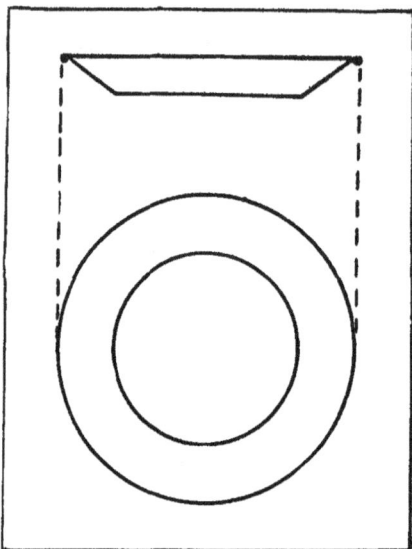

Figure 7. Plan and side view of gold miner's pan.

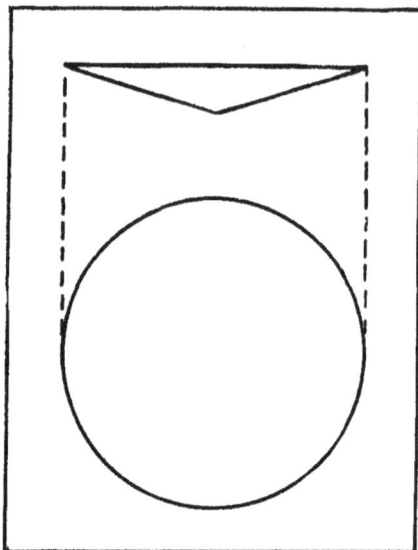

Figure 8 Plan and side view of batea.

but the coating chips off. Copper-bottomed pans with steel sides and solid copper pans with the inside copper surface coated with mercury (quicksilver) are often used to collect gold by amalgamation. Diameters of pans at the top vary from ten inches to eighteen inches with depths of from two to three inches. The inside surface of the pan must be kept smooth and free from oil or grease.

Batea.

The batea, Figure 8, is used for panning gravel in Latin America and in the Asiatic countries. It is usually made from hard wood, but sometimes sheet metal is used. Top diameters run from fifteen to thirty inches. A batea is handled in much the same way as a miner's pan, but the concentrates (gold and the heavy minerals) collect near its center.

Miner's spoon.

The miner's gold washing spoon, Figure 9, is sometimes used to test or sample small quantities of sand or dirt. It is made

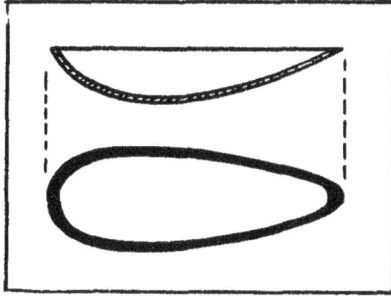

Figure 9. Plan and side view of miner's spoon.

from either ordinary horn, hard rubber, copper, or polished steel. Spoons made from horn and hard rubber break easily, steel spoons rust, and in a dry climate, horn spoons check badly.

Other panning utensils.

Frying pans, pie plates, and similarly shaped utensils are also used to pan gravel.

Panning.

Panning is an operation that is very difficult to describe and one that no two persons do exactly alike. It is best learned by observation supplemented by advice and practice. No matter whether a pan, batea, or spoon is used, the operation is essentially the same, but it is assumed that a pan is being used in the attempt to describe the procedure, which follows:

Fill the pan nearly full of dirt and place it in water deep enough to cover the pan and its contents. Work over the contents with both hand, breaking up the lumps and throwing out the stones. After the contents have been thoroughly disintegrated and the stones have been removed, grasp the pan with both hands, at opposite sides of the top, for the panning operation. Holding the pan about level, give it a rotating motion, rapidly alternating the direction, so as to agitate the contents and allow the heavy particles to settle to the bottom. Then move the hands until they are a little back of the middle of the pan. This action tips the pan away from the panner. With the pan in this inclined position, give it a circular, sidewise, shaking motion that washes the contents from side to side. This brings the lighter material to the surface and washes it toward the front or lip of the pan while the heavy particles work their way toward or remain on the bottom. Some of the lighter material washes out of the pan. To remove more of the lighter material, cause water to flow over it by raising and lowering the lip of the pan through the surface of the water. An experienced panner usually scrapes off considerable of this light material with his thumb. These operations are repeated until nothing but the concentrates (gold and the heavy minerals, called black sand)[4] are left in the pan.

Cleaning concentrates.

The concentrates are saved until a fair amount has been accumulated. Carefully panning the accumulation will considerably reduce the quantity. Removing the magnetic particles (chiefly magnetic iron) by use of a magnet will reduce the bulk still more. The magnet works best when the concentrates are dry. Covering the ends of the magnet with paper or cellophane helps to keep the magnet clean since, on withdrawing the magnet from the covering, the magnetic particles drop. Careful blowing, especially with a blow box, will also reduce the proportion of worthless stuff. The larger gold colors can be picked out with a sharp-pointed pair of tweezers. Fine or flour gold may be collected by amalgamation.

Amalgamation.

Clean mercury, if ground or agitated with the concentrates, will catch the gold. To accomplish this result, some miners grind the mercury and concentrates in a mortar or muller; some agitate and grind the mixture with a couple of black iron slugs in a pan; others agitate the mixture in a bottle, and still others use a copper amalgamating pan. For cleaning purposes, it is advisable to add about a teaspoonful of lye, a little dilute nitric acid or a little weak cyanide solution to the material.

Many placer miners add a teaspoonful or so of mercury to a pan of dirt to catch the flour gold, especially when flour gold predominates. The mercury is usually added after the coarsest gravel has been removed.

Copper amalgamating pans.

When solid copper or copper-bottomed, steel-sided pans are used for amalgamating the gold, the mercury is not only added to the material, but the inside copper surface of the pan is coated with either metallic mercury or silver amalgam.

Instructions for amalgamating copper pans.

The first part of one method for amalgamating the pan is to

---

[4] The heavy minerals in concentrates, commonly called black sand, may contain magnetite (magnetic iron oxide), ilmenite (iron-titanium oxide), hematite (non-magnetic iron oxide), iron pyrites (iron sulfides), marcasite (white iron sulfides), rutile (titanium oxide), wolframite (iron-manganese tungstate), zircon (zirconium silicate), garnet, and other heavy minerals.

The specific gravities of these minerals are as follows: Magnetite, 5.1; ilmenite, 4.7; hematite, 4.8; iron pyrites, 4.9; marcasite, 4.9; rutile, 4.2; wolframite, 7.2; zircon, 4.7; and garnet from 3.5 to 4.5. So rarely does any mineral in the concentrates, except gold, have value that a prospector is running practically no risk of overlooking something worth saving if he assumes that, in this state, his "black sand" is worthless.

thoroughly clean and brighten the copper surface. Some use one thing and others use another to accomplish this result, but plenty of "elbow grease" is essential. Weak cyanide solution (very poisonous), weak nitric acid, lye, very fine emery, very fine sand and even Dutch Cleanser or Sapolio, when rubbed over the copper, clean and brighten the surface. After the copper is clean and bright, mercury is sprinkled on it and rubbed in with a piece of canvas, blanket, or toweling. Good mercury shakers may be made from a small bottle with a large mouth over which muslin or canvas is stretched or from a short half-inch black iron pipe nipple with one end capped and muslin or canvas stretched over the other end. A damp mixture of ten parts fine sand, one part sal-ammoniac, and a little clean mercury, when rubbed over the clean, bright copper surface with a piece of canvas or blanket, produces a uniformly silvery surface.

Another method for amalgamating the copper surface of the pan is to pour into the pan a dilute solution of mercury. This solution can be made by dissolving liquid mercury in nitric acid. Dilute this acid solution of mercury with rain or distilled water to about ten or fifteen times its original volume. Pour this into the pan and, in a few minutes, the copper will be coated with mercury. More mercury can then be rubbed in.

Mercuric chloride, when dissolved in water and poured into the copper pan, gives a good coating of mercury.

Panning with an amalgamated pan.

Panning with an amalgamated pan is done in just the same way as when mercury is not used. After the gold has been caught by the mercury and the worthless minerals have been washed out of the pan, the amalgam is scraped up (use a putty knife or hard rubber scraper) and collected . The excess mercury in the amalgam is then squeezed through a soft, damp chamois or canvas, leaving hard amalgam in the chamois. The gold is separated from this amalgam by methods outlined on pages 115-119 of this bulletin.

Quantity of gravel that can be panned in eight hours.

The amount of gravel that can be panned in eight hours by a fairly good panner is about fifty pans or about one quarter of a cubic yard. One cubic yard of average gravel in eight hours is about the maximum that can be carefully panned by a very skilled panner when all conditions are favorable. If the gravel is cemented or if sticky clay is present, this quantity is reduced.

ROCKER (CRADLE)

Next to the pan, the miner's rocker (cradle), Plate 11 and Figure 10, is the commonest machine used in placer mining. If gold values can be profitably recovered with a pan or batea, rockers can be advantageously used since they handle somewhat larger amounts of dirt.

Capacity.

With a fair-sized rocker, two men working together can wash from two to four cubic yards of average dirt in eight hours, but sticky clay and cemented gravel lower the capacity. Clayey dirt should be soaked and stirred up in a puddling box before it is put through the rocker or, otherwise, the clay (gold robber) may carry off some of the gold. The rocker catches coarse gold effectively, but flaky, porous, or flour gold is apt to float off with the tailings.

Construction Details.

The miner's rocker (cradle), made in many sizes and shapes (no two of which are absolutely identical) consists of a hopper (screen-box), an apron, and a riffled box or trough which is open at one end. This trough is mounted on two rockers which are set crosswise beneath the bottom. Other features are usually incorporated in its construction but the ones mentioned are the essentials.

Rockers are made from ten to 45 inches in height with an outside width of from eight to thirty inches. Lengths range from fifteen inches to eight feet or longer. Short rockers are poor flour gold catchers, long ones catch flour gold better, but are difficult to transport, while extra high rockers make the shoveling and other work too laborious. The usual length is from three to five feet.

Soft lumber, which will not shred or rough up under working conditions, is the best to use in the construction of the body of the rocker. The bottom should be made from a single board, free from knots and cracks, with the top side planed. When such lumber is not available, the bottom is covered with a piece of carpet, canvas, burlap, blanket, corduroy, corrugated rubber, cocoa matting, hide with the hair on it, or similar material over which the riffles are placed. Such a covering acts as a good flour gold catcher and it is, therefore, a good idea to use it in all rockers. This covering should be taken up occasionally and cleaned of its gold (washed in a tub or burned and the gold panned from the ashes).

Apron.

The apron is a canvas-covered framework made of about ¾x1½ inch lumber. The side pieces are usually extended at the lower ends a little beyond the lower crosspiece, so as to allow clearance for the dirt. The canvas is tacked onto the framework so as to leave a sort of sag or pocket about an inch deep, at the lower end. Instead of the sag pocket, some persons fold back the canvas and make one or more cross pleats that act as riffles and pockets. The material that passes through the screen of the hopper falls onto the apron where some of the coarse gold and other heavy minerals are caught, the balance washing over the end of the apron onto the head end of the rocker. The apron should fit loosely enough so that it can be readily removed and cleaned after several batches have been worked through the rocker. The concentrates from it are accumulated and washed in a pan.

Riffles.

Riffles, as shown in Figure 10, are used to catch the gold that washes over the apron. When the dirt carries so much black sand that its banking up behind the riffles prevents the riffles from catching the gold, some, if not all, of them are dispensed with and a covering of carpet, blanket, burlap, cocoa matting, or similar material is used to catch the gold. Wire cloth, wooden cleats, or metal strips, tacked down over the covering hold it in place and act as riffles.

Hopper or screen box.

A typical hopper (screen box) is shown in Figure 10. The ends should fit loosely, but not too loosely, in the rocker. About half an inch of clearance between the sides of the rocker and the sides of the hopper gives the proper bump. The cleats that support it should be set so that the bottom of the hopper is nearly level when the rocker is set up on its bases. The screen or bottom of the hopper is usually made from thin sheet metal, (about 18-guage iron). Some use the tin from a five-gallon oil can. This is perforated with holes about half an inch in diameter, spaced about two inches apart. Some do not punch the holes in the sheet metal of the hopper screen right up to the ends of the hopper but leave from three to four inches of solid sheet metal at one or both ends. This is done so that all of the fine stuff that passes through the screen will fall onto the apron for a preliminary concentration. Some operators use half inch or larger wire screen cloth for this purpose but it does not work very well.

Slope of bottom.

The slope or grade for the bottom or floor of the rocker depends upon the character of the material being handled and is determined by the "cut and try" method. Light material requires a flatter slope than gravel that contains much heavy minerals. The usual practice is to give the bottom a slope of about two inches in three feet, which can easily be done by setting one base plank higher than the other. Some people make one of the rockers about two inches higher than the other rocker and use level bases. A heavy plank is usually used as a base under each rocker, and these planks should be well secured so that they will not move and shift around when the rocker is in operation. Cross pieces nailed from one plank to the other make the base more rigid. A hole or groove in each base (plank) must be provided to take care of the spikes that prevent the rocker from working down grade. Cleats fastened to the lower side of each plank are sometimes used instead of spikes.

Amalgamation.

Mercury is sometimes used in rockers, particularly to catch the flour gold that otherwise washes over the riffles and out of the rocker with the tailings. Many placer miners pour a little mercury behind the riffles and some of them use an amalgamated plate, also. When added to the riffles, care should be taken to pour the mercury so that it does not break up into small globules that will wash over the riffles and out with the tailings. Some placer dirts contain minerals that flour the mercury and with them, it may not pay to use mercury.

Operation.

Enough material is dumped into the hopper (screen box) to fill it from one-half to two-thirds full. While a stream of water is poured over the material, usually from a dipper, the rocker is given a rocking motion and kept rocking. Small rockers are usually rocked by hand but motive power is being used more and more for this purpose. The material is worked over and the clean stones and boulders are picked out, inspected, and, if found valueless, discarded. The water washes the fine material through the screen onto the apron where some of the gold and heavy minerals are caught. The material that does not remain on the apron washes over its end onto the bottom of the rocker where more of the gold and heavy minerals are caught behind the riffles. The lighter, worthless material washes over the riffles and out of the rocker. It is advisable occasionally to test, by panning, some of the tailings (waste) flowing out of the rocker to find out if gold values are being lost.

Water.

The water should be added in a steady stream and the volume should be sufficient to carry the waste out over the riffles without banking. Too much water may wash the gold out of the rocker with the waste, while too little, especially if much clay and black sands are present, will not allow the gold to settle and be caught. More water is usually used in rocking than in panning, the amount varying much. If used sparingly, especially if some of it is reclaimed by the use of a settling basin and used over again, the amount of water needed varies from fifty to one hundred and fifty gallons or from two to six barrels of water for each cubic yard of dirt put through the rocker. Some find it more advantageous to haul the gravel to the water than the reverse.

Clean-ups.

The frequency of clean-ups depends upon the amount of gold being caught and how the rocker is functioning. It is advisable to watch the concentrates behind the riffles and govern the frequency accordingly. The apron, where most of the coarse gold

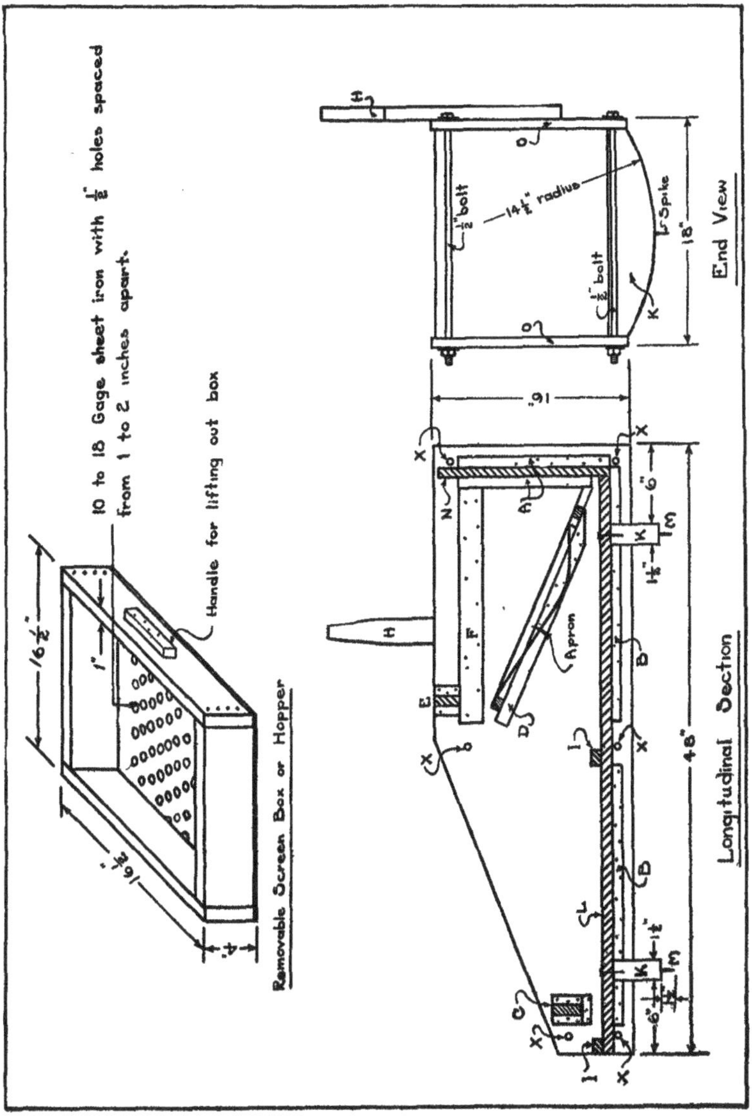

Figure 10. Knockdown rocker

values are caught, should be withdrawn and cleaned of the accumulated concentrates after five or ten batches of dirt have been put through the rocker. Less frequently, once or twice in an eight-hour shift, the riffles are cleaned of the concentrates that have collected behind them, but the richness of the concentrates governs these factors. After a quantity of these concentrates has accumulated, they are cleaned and the gold is separated from the worthless stuff as already outlined under panning.

Construction details for knock-down rocker.

Figure 10 shows a knock-down rocker and is taken from the article entitled "How to Make a Rocker" by W. H. Storms in the June 24, 1911, issue of the Engineering and Mining Journal. A longitudinal section through the center of the rocker, an end view, and a hopper (screen box) are shown.

A—Cleats—The back (N) slips in between them.

B—Cleats—To hold bottom (L) of rocker.

C—Cleats—To hold front crosspiece.

D—Cleats—To support canvas apron.

E—Cleats—To hold top crosspiece.

F—Cleats—To support hopper (screen box). They should be placed so that the bottom of the hopper is about level when the rocker is set up on its bases.

X—Bolt holes for ½-inch iron bolts used in holding rocker together.

I—Riffles—¾-inch high by 1 inch wide.

H—Handle for rocking rocker. (Some rockers have the handle fastended to the hopper.)

K—Rockers.

L—Bottom board of rocker (1-inch lumber dressed to ¾-inch is heavy enough).

M—Spike to prevent rocker from slipping down grade.

N—Back of rocker.

O—Sides of rocker.

THE LONG TOM

The long tom, Figure 11, is a modified sluice box which is often used in place of a rocker. Dimensions vary greatly, but usually range from six to twelve feet in length, the upper end being from fifteen to twenty inches wide, the lower end being from 24 to 32 inches wide, with sides from six to twelve inches high. Attached to the lower end is an inclined screen (B) set at an angle of 45 degrees to the bottom. This screen is a piece of heavy sheet iron perforated with three-eights or one-half inch holes. The tom is usually given a slope of about one inch per foot of length. A wide riffle box (C), usually set on a flatter grade than the tom (A), with its upper end set under the lower end of the tom, receives the fine material and water that pass through the holes in the screen (B).

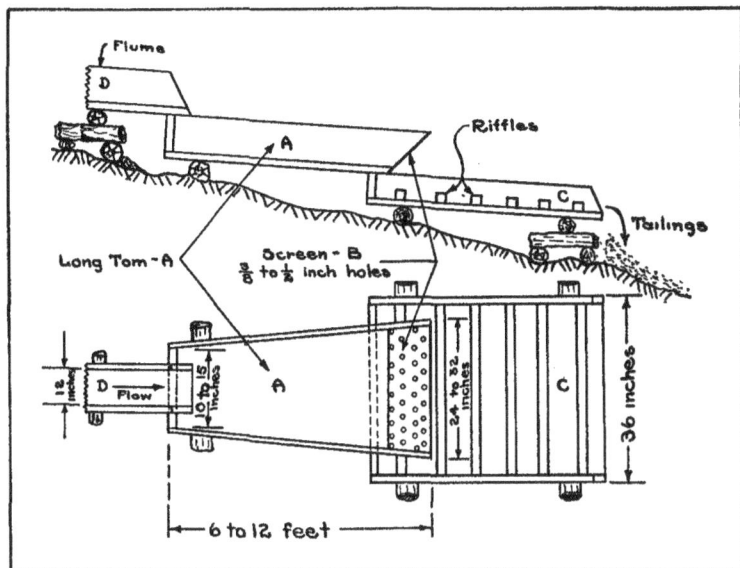

Figure 11. Long Tom.

Operation.

The material is shoveled into the tom (A), at the head end or into the flume (D). A stream of water flows from the flume (D) or a pipe onto the material which is worked over by means of a rake, fork, or square-ended shovel so as to break up the lumps of clay and to clean the dirt off the stones, the clean stones and boulders being forked or shoveled out. The fine material is worked through the holes in the screen and falls into the riffle box where the gold, with or without the aid of mercury (quicksilver), and the black sand settle behind the riffles. The tailings or light, waste material wash out of the riffle box. Often the riffle box is supplemented and followed[5] by one or more sluice boxes, the bottoms of which are covered with canvas, carpet, fleece, burlap, or some similar material for catching the fine, flour gold that passes over the riffles.

Toms are regularly cleaned up, the gold and amalgam collected from the riffles being washed in rockers or pans. The amount of material that can be handled by a tom in eight hours, two

---

[5] An inclined screen allows the pebbles and coarse sand to be easily discarded. To get best results the pulp (fine sands, flour gold, and water) should flow over the mats or covering in a rough shallow stream

Plate 12.　Small sluice in operation, Chase Creek.

men working together, is as high as five[6] cubic yards of average dirt and from two to three cubic yards of somewhat cemented material. The long tom requires an ample water supply, but it uses less water than the sluice. In small-scale operations, where lumber is expensive and scarce, it finds favor. Long toms are now little used in this country because, where the grades are satisfactory and an ample supply of running water is available, the sluice is usually as effective and requires less labor to operate.

SLUICES

Sluice is a term applied to any sloping trough or ditch that is used by placer miners for the purpose of separating and catching gold from the placer dirt that they wash through it. A ditch, cut in rock or dug in hard gravel—the irregularities in its bottom acting as riffles—that is used for this purpose is called a ground sluice. A trough mode from wood, with riffles along the bottom, is called a box-sluice. Box-sluices are commonly made up of 12-foot sections, called sluice boxes, usually butted together and held in place by wooden strips. When the head boxes in a sluice have to be frequently moved, telescoping sections are sometimes used. Due to unfavorable water supplies, few, if any, sluices in Arizona are over fifty feet long while most sluices that are used in this state are from six to twelve feet long and from six to fourteen inches wide.

Sluicing, when applied to placer mining, is a method of separating and catching placer gold from placer dirt by the use of a sluice and running water. The running water washes the lighter waste material through and out of the sluice, thereby performing much of the work done in panning and rocking. The gold and heavy minerals settle and are caught behind the riffles.

Much coarser material is usually put through sluices than through rockers. The large boulders are usually screened off by the use of a grizzly or are forked out. The stones as they are washed through the sluice tend to grind and brighten the gold, making it more amenable to amalgamation. They also disintegrate the dirt so that any enclosed gold particles are liberated. Where a small sluice, as shown in Plate 12 or Figure 13 is used, the feed is often screened before it is fed to the sluice. When used under favorable conditions, the sluice handles placer dirt at minimum costs.

Riffles.

In placer mining, riffles are obstructions along the bottom of a sluice or rocker that retard the progress of the heavier minerals and form pockets to catch the gold.

Vizetelly[7] defines the term "RIFFLES," when used in mining,

[6] Wilson, E. B., Hydraulic and Placer Mining, 3rd ed., John Wiley & Sons, Inc., New York, 1918, p. 70.

[7] Vizetelly, F. H., The College Standard Dictionary of the English Language, Funk & Wagnalls Company, New York, 1932, p. 980.

as follows: "1. A groove or indentation set in the bottom of an inclined trough or sluice, for arresting gold contained in sands and gravels. 2. A cross-slat or cleat rising above the bottom of such a sluice and adapted for catching gold."

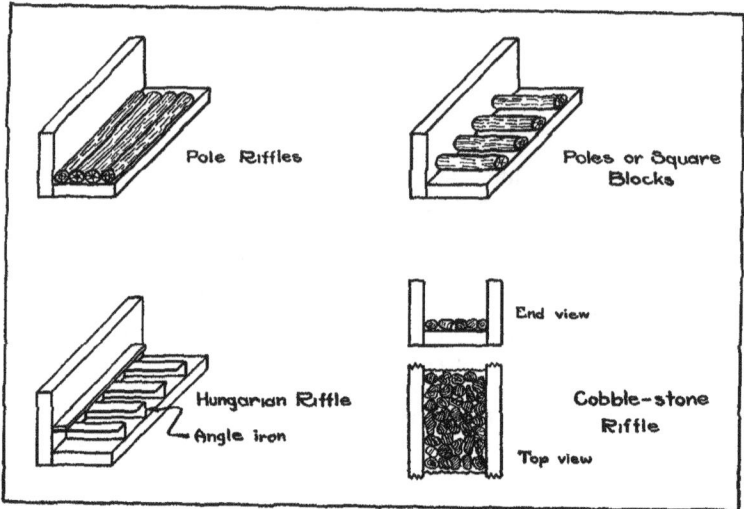

Figure 12. Various types of riffles.

"Riffles[8] in placer mining have three chief functions: (a) to retard material moving over them and give it a chance to settle; (b) to form pockets to retain the gold which settles into them; (c) to form eddies which roughly classify the material in the riffle spaces. Their exact operation is not well understood. The strength and shape of eddies (the "boil" of the riffle) is affected by the shape and spacing of riffles, their position with respect to the direction of flow and the velocity of current. The boil must be strong enough to prevent the riffles from filling up with heavy sand (packing) and not too strong to prevent the lodgment of gold."

There are many different kinds and shapes of riffles, made from various materials, a few in common use being shown in Figure 12, the riffle used being selected by the placer miner from the material available. Riffles made from wood, iron, steel, or cobble stones are used for the ordinary run of mixed coarse and fine gravel, while carpet, burlap, blanket, canvas, cocoa matting, hides with the hair side up, corduroy, corrugated rubber, and such materials, sometimes tacked and held down by wire cloth, wooden cleats, or metal strips, are commonly used for fine sands

---

[8] Peele, R., Mining Engineers' Handbook, 2nd ed., John Wiley & Sons, Inc., New York, 1927, p. 915.

and to catch flour gold. When the dirt carries so much black sand that it banks up or packs behind the riffles and prevents them from catching the gold, some, if not all of the riffles are dispensed with and the above mentioned materials are used to catch the gold.

Slope.

. "The slope[9] of the sluice depends upon the character of the gravel and gold, the kind of riffles used, and the quantity of water available." A limited water supply makes it necessary to use a steep slope if the largest possible quantity of gravel is to be run through the sluice. Moderately fine gold is caught best on a steep slope when the water is spread out in a rather shallow, rough stream. Slopes given sluices vary from four to eighteen inches for each twelve feet length, the usual slope being about six inches for each twelve feet of length.

Water consumption and capacity.

More water is required in sluicing gravel than in panning or rocking and, unless an ample water supply is available, it is useless to build a sluice. The water supply should be sufficient to furnish all of the water needed of efficient sluicing while the sluice is in operation, and varies greatly, running from ten to fifty cubic feet of water for each cubic foot of gravel put through the sluice. The following table[10] gives data on the capacity of two sluices:

Width, inches	Depth of flow, inches	Grade, per cent	Water flow, cubit feet per minute	Cubic yards of gravel per 24 hours
10 - 12	6 to 7	4.16	45	65 to 135
12 - 14	10	6.2	100	150 to 300

Amalgamation.

Coarse gold is usually caught in the upper riffles of a sluice, but the flour, porous, or flaky gold tends to float in the moving water and wash away with the waste. To catch this flour, porous, or flaky gold, mercury is sometimes added to the riffles. Care must be taken in adding the mercury to prevent it breaking up into small globules that will wash away.

Clean-ups.

Clean-ups depend upon the amount of gold being caught and take place at more or less regular intervals. After the lighter waste material has been more or less washed out, the riffles are removed and the concentrates are then carefully collected and cleaned up in a rocker or pan.

---

[9] Taggart, A. F., Handbook of Ore Dressing, John Wiley & Sons, Inc., New York, 1927, p. 645
[10] Young, George J., work cited, p. 434.

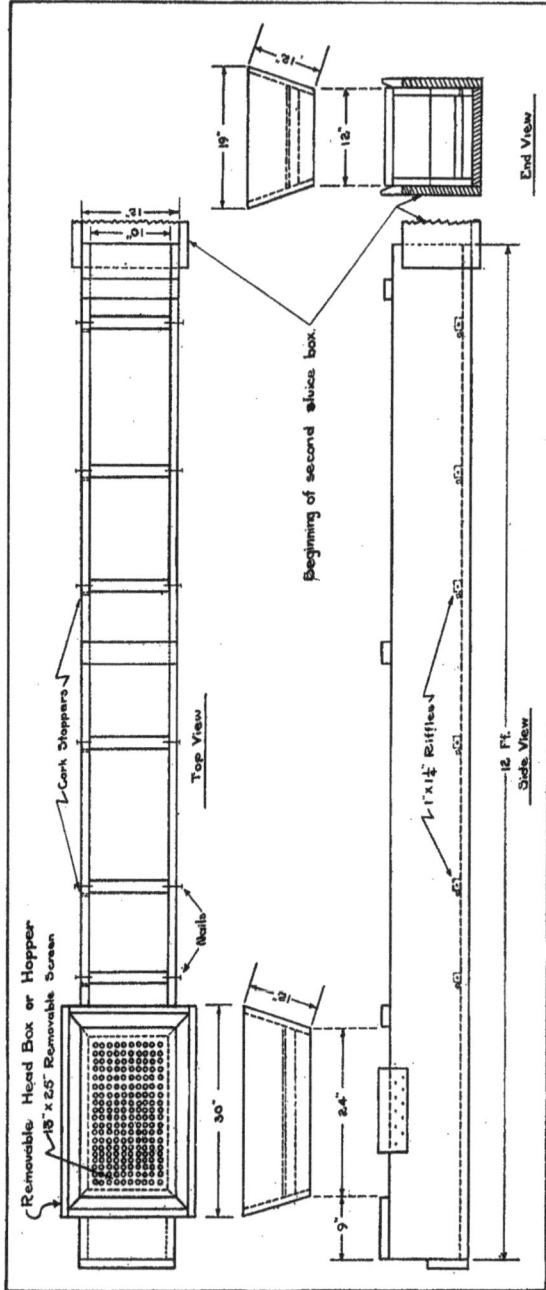

Figure 13. Details of small sluice-box.

Small Sluices.

Figure 13 shows three views of a small sluice box, Figure 14 shows a typical sluice lay-out, and Plate 12 shows a small sluice box in operation. Although sluice boxes are often made from rough lumber, such boxes are much harder to clean up than if the inside surfaces of the boards are planed. The joints should be calked water-tight with lamp wicking or other calking material because fine gold will go where any water leakage occurs. Wooden riffles can be fastened by nails driven into their ends through the sides of the box. Since the riffles are removed when cleaning up the sluice, it is a good idea not to drive the heads of these nails all the way in as they can then be easily drawn out and the riffles lifted. Wedges are sometimes used to hold the riffles in place but they do not work very well. As already mentioned, part or all of the bottom of the sluice box, especially if the gold is fine and much black sand is present, may advantageously be covered with burlap, blanket, carpet, or similar material. At the clean-up, the covering is taken up and washed in a tub to recover the gold. When either the covering or the sluice box is worn out, it is burned, and the ashes panned for any gold that is in the cracks of the wood or in the meshes of the covering.

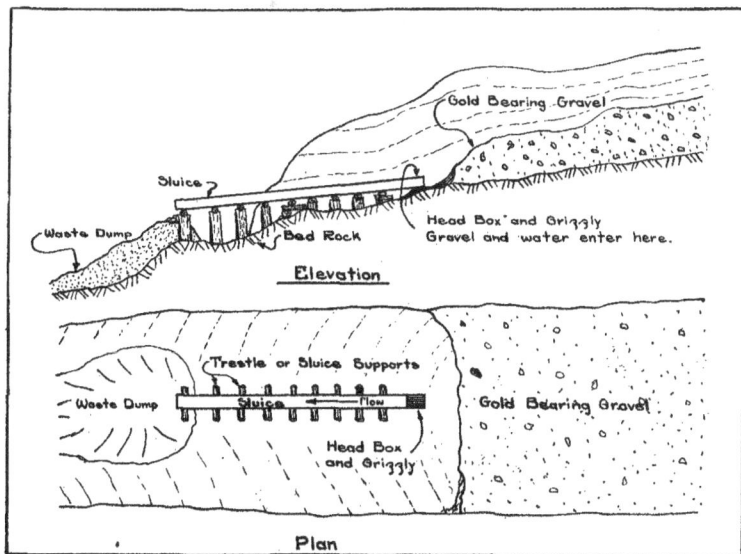

Figure 14. Sluice-box lay-out.

Wet Methods vs. Dry Methods

All the foregoing methods of wet gravel treatment necessitate the presence of ample water. The methods are so far superior in efficiency to the dry methods briefly described in the following paragraphs that by "hook or crook" wet methods should be used. In Arizona, except perhaps in the driest of the desert regions, every dry arroyo or canyon will run water at least once a year, even if for only a few hours. If pay dirt is found in such an arroyo or canyon in sufficient quantity, the building of earth-work dams and ditches above the deposit during the dry seaon to catch and impound the one or two rainy season cloulbursts affecting the area may permit the rapid working of the deposit with ample water, and at a far greater profit than by the laborious and costly hauling or pumping of water to the deposit or the hauling of the gravel to the nearest permanent water. In this work, the enormous force of running water should be realized, and reservoir sites should be chosen amply large to protect the dams from flood water. The "farming" of several such deposits might be possible in the dry season by a prospecting party of several men, with the expenditure of no funds other than a grub-stake to tide them over until the one or two rains fill their reservoirs.

Dry Concentration

The recovery of gold from gold-bearing gravel in arid districts, where water is scarce and too expensive to be profitably used in any of the small-scale wet concentrating methods already outlined, has been responsible for the development of many ingenious methods and machines. Practically all of the methods and machines that are used in the field to accomplish this purpose utilize moving air or wind instead of water as a medium of separation.

From tests made at the University of Arizona[11] the following facts have been obtained.

"A dry concentrator will not make as high recovery as a wet concentrator. Under favorable conditions, the recovery will be approximately ten to fifteen per cent less with a dry machine as compared with a wet machine. It follows, therefore, that a wet machine should be used in preference to a dry machine where water is available.

The difficulty with the practical operation of dry concentrators is due to the fact that they require the material treated to be in an ideal condition. First, the material must be dry; moist or damp material is not satisfactory as feed for dry machines. Second, the material must be disintegrated and this condition

---

[11] Chapman, Dr. T. G., personal communication, 1932.

practically limits the use of dry concentrators to sandy, dry material or material that can be sun-dried and easily disintegrated."

When Dr. Chapman refers to disintegrated material, he means sand, gravel, etc., in which the particles of gold are free—unattached to and not included within the waste material. If some clay is present, it may become very hard when dry and cement the gold to the sand or gravel which may still appear to be loose or disintegrated. Such gold is lost in dry-washers, but might be released and recovered if treated with water.

A third drawback to the use of dry-washers is the fact that all nuggets too large to pass through the screen, in which the openings should be relatively small since close sizing is desirable, are lost unless the material that collects on the screen is examined very carefully. A few years ago, a $152 nugget was in this way left lying on the waste dump in the Weaver district (see Plate 3.)

DRY CONCENTRATORS

Blanket.

One of the primitive methods involves the use of a dry blanket with which the dry, gold-bearing material is tossed up into a strong wind. The wind winnows it and blows the light fines away. The coarse stuff is picked out by hand and the fine concentrates remaining on the blanket are then treated by blowing and hand picking until the gold is collected. Some of the gold is caught in the hair of the blanket.

Dry panning and blowing.

The dry, gold-bearing gravel is dumped into a pan and shaken up so as to bring the lumps and coarse stuff to the top. After they have been removed, the remainder is slowly poured, from about shoulder height, into a second pan which is placed on the ground. A strong wind blowing through this stream of material winnows it and carries away the light fines. This operation is repeated several times until a concentration has taken place. The concentrates are then winnowed by tossing them up from the pan into the wind. Following this operation, the material remaining is panned just as in water. The concentrates from this panning are then cleaned further by blowing with the mouth.

Dry-washers.

Practically all of the small-scale, dry concentrating machines that are used in the field are dry jigs although dry tables are being tried out on one or more large-scale developments. Dry jigs, locally called dry-washers, differ widely in design and construction, but practically all of them are built so as to subject a bed of the gold-bearing gravel to intermittent pulsations of air. These blasts of air bring the light particles to the top, the gold and heavier paricles settling beneath.

Plate 13. Dry-washer.

Plate 13 shows a dry-washer that is made in this state. Like others of its type, it consists essentially of a screened hopper and feed-box and a cloth-bottomed, inclined tray with cross riffles, beneath which is a bellows. The bellows forces intermittent blasts of air up through the cloth. This action agitates the material, brings the lighter fines to the top, and blows them away. The gravel is fed through the hopper upon the upper end of the tray and is slowly moved down the slope by this agitation. The gold lodges behind the upper riffles and the material of lower specific gravity (less weight) flows over the riffles and gradually passes out of the tray at its lower end. To catch the flour gold, some pour a little mercury behind the riffles.

A machine of this type is usually operated by two men. One turns the wheel that operates the bellows while the other feeds the gravel and watches its progress over the riffles. Some dry-washers use motive power to operate the bellows. When the riffles apear to be loaded with concentrates, the tray is removed and the concentrates are transferred to a pan for further cleaning and concentration.

The capacities of dry-washers range in eight hours from a couple of cubic yards of dry gravel upward. When properly operated, they catch rather effectively the coarser gold that passes the screen, but fine, flaky, flour gold is apt to go off with the waste and be lost.

Detailed diagrams of dry-washers are not included in this bulletin since it appears to be undesirable to recommend any particular type or types. Almost any standard style of machine will work fairly satisfactorily where conditions are all favorable, but such conditions are so rarely encountered in Arizona that the use of dry-washers is very apt to prove disappointing.

Separating Gold From Gold Amalgam

The gold amalgam that is produced by any of these methods always contains an excess of mercury. Some of this mercury can be removed by hanging the amalgam up in a muslin or canvas sack and allowing the excess mercury to drain off into a pan. Squeezing it through canvas or a damp chamois gets rid of more. The gold that is in the hard, dry amalgam that is left in the chamois or canvas can then be separated from the mercury either by driving the mercury off by heat or by dissolving the mercury in nitric acid. The gold is left in the residue when either process is used.

Retorting.

The separation of mercury from amalgam by the use of heat (distillation) is commonly called retorting.

Retorting of gold amalgam is accomplished by heating the amalgam to a temperature that is high enough to vaporize and drive off the mercury but not the gold, the gold remaining as a residue.   This result can be accomplished because the normal boiling point of mercury is about 675 degrees Fahrenheit while that of gold is about 4720 degrees Fahrenheit.

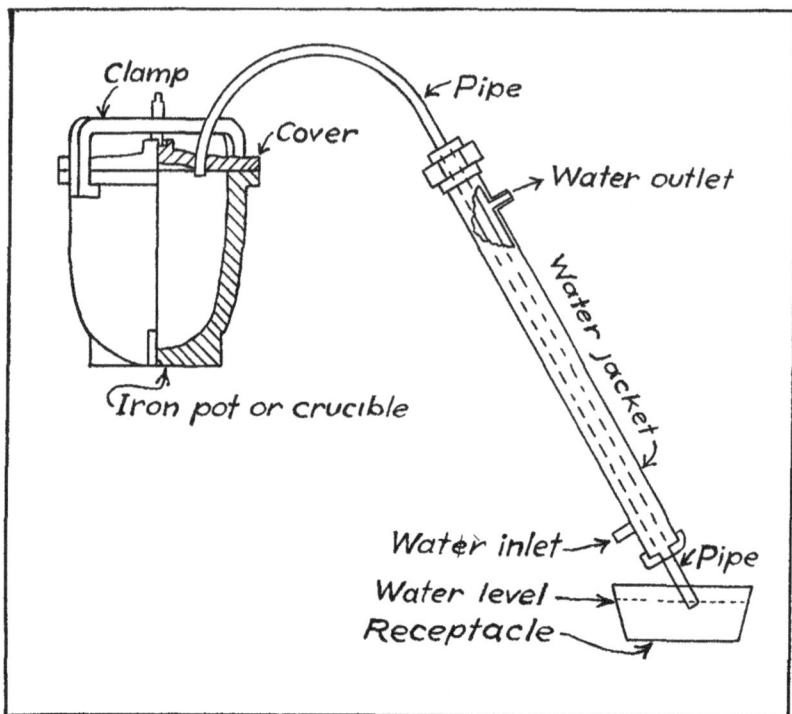

Figure 15.   Retort.

Retorts.

Figure 15 shows a retort that is much used for retorting small batches of amalgam.   In the upper left part of this figure is an iron crucible or pot into which the amalgam is charged.   On top of this crucible is a tight fitting cover that is held in place with a clamp.   Into a threaded hole in the cover is screwed a bent iron pipe which removes the hot mercury fumes that are driven or distilled off from the amalgam.   This outlet pipe is about four feet long and is bent so that most of its slopes downward when the retort is set up.   This slope permits the mercury to run down and out of the pipe after it condenses to a liquid.   This sloping part of the pipe is kept as cold as possibly in order to cool the enclosed hot mercury fumes so that they will condense to liquid mercury.   In the figure, a metal-jacketed outlet or condenser pipe is shown.   Cold water passes through this jacket continuously and cools the pipe during the retorting operation.   A home-made retort, Plate 14, usually has the outlet pipe wrapped with burlap

Plate 14. Retort (home-made).

or some other loose material over which cold water is poured. The discharge end of the condenser pipe of a retort should barely dip beneath the surface of some water in the receptacle that catches the retorted mercury. The water will condense all of the mercury fumes that escape from the pipe but only a little water will be drawn up into the pipe when the retort cools. If the water is too deep over the discharge end, the vacuum produced in the crucible of the retort when it is cooling may suck water up into the hot crucible, make steam and cause an explosion. To overcome this danger, some wrap a wet cloth around the discharge end of the pipe in such a way that the cloth acts as a conduit and carries the liquid mercury into the water of the receiving pan.

The purified liquid mercury from the retort is suitable for further uses.

Plate 14 shows a home-made retort, made from an iron (black) pipe nipple, pipe caps and a bent iron pipe.

Charging an amalgam retort.

The inside of the pot or crucible is thoroughly cleaned and is then painted or coated with a thin emulsion of clay, chalk or wood ashes. This is done to prevent the gold residue that is left in the crucible, after the mercury has been distilled or driven off, from sticking to the inside of the crucible. Lining the inside of the crucible with a double thickness of newspaper or wrapping each lump of amalgam in newspaper helps to prevent this sticking. The amalgam, in small lumps or balls (not over an inch in diameter), is then put into the crucible and a few pinches of powdered charcoal or bits of paper are sprinkled over the charge. The cover is then fastened to the pot or iron crucible. To prevent the poisonous mercury fumes from escaping through any crack, all connections in the retort should be luted (sealed up) with thick, wet clay, chalk, or wood ashes.

Heat swells amalgam. The retort should not be filled over one-half full of amalgam because, if too large a charge is used, the amalgam, on being heated, will swell and may close the outlet pipe thereby causing the retort to explode. The retort should be heated rather slowly at first, the heat being gradually raised to a dark-red. Near the end of the operation, the heat may be safely raised, for a few minutes, to a cherry-red. Too quick heating will liquify and boil some of the mercury. This boiling mercury may throw pieces of the amalgam up into the outlet pipe, stop it up, and cause an explosion. Due to the great weight of the amalgam, too high a heat over too long a time may cause the red hot amalgam pot or crucible to bulge. Retorting a small batch of gold amalgam should not take more than from two to four hours. The crucible should be allowed to cool off before it is opened.

Potato method.

An ingenious and simple method to use in retorting a small quantity of amalgam (up to an ounce in size) is by means of a potato. Choose a large, well-rounded white potato, fresh and without drying cracks, and cut it in two halves. Scoop out, in one half, a hole large enough to hold the amalgam and place the amalgam in the hole. Join the two halves and wire them tightly together Place the potato in the hot ashes of a camp fire and let it bake done (from a half to one hour). Remove the potato, let it cool, unwire it, and a gold button will remain in the center. After removing the button, place the potato in a pan and squeeze out the distilled quicksilver from the pulp. **Do not eat the potato.**

A modification of this potato method is to use one-half of a white potato to cover the ball of amalgam placed in an indentation in the bottom of a frying pan or blade of a shovel.

Another method is to roll the ball of amalgam in paper and place it on the blade of a shovel or in a frying pan that has been painted with thick clay. The trouble with this last method is that, upon being heated, the mercury and amalgam spit and shoot, with certain loss of the gold values. All of these last mentioned methods lose most, if not all, of the mercury.

When small amounts of amalgam are to be retorted, it is a good idea to send them to an assayer and have the work done by an experienced man who has the proper equipment. Retorting and burning the mercury from amalgam is a very dangerous operation which should *never* be conducted in a closed room. The fumes of volatilized mercury *are very poisonous.* If a human being *inhales them, he may lose his teeth* or *he may lose his life.*

Nitric acid method.

The gold can be separated from the amalgam by hot dilute nitric acid (one part strong acid to about two parts rain or distilled water). Dilute nitric acid does not dissolve gold, but it does dissolve mercury and silver. After the mercury and silver have gone into solution, the solution is poured off from the gold residue.

This nitric acid solution, which contains the mercury, can be used to amalgamate copper pans. Most placer miners who use this process of separation throw the solution away even though they know that it contains silver and mercury which can be saved by precipitating them on metallic copper.

GOLD FLOTATION

The flotation process.

Within the past generation and especially within the last twenty years, the flotation process has revolutionized the milling methods used in the extraction of most metals from their ores. Tremendous tonnages of copper, lead, zinc, gold and other ores, whose values were too low grade or for some other reason could not be worked at a profit by the other and older methods of extraction, have, when subjected to flotation, returned good profits to mine operators.

Discoveries and improvements, both in mechanical equipment and reagents, have been made with unusual rapidity. The operators in this industry have become much more proficient in their work. Its rapid adoption by the mining industry may be appreciated when one realizes that, until 1910, this method of extraction had been used on only a few tons of ore, while, in the year 1919, alone, more than 29 million tons of ore were treated by flotation. The speed with which this new and revolutionary method of extraction was adopted and put into practice by the mining industry has been truly remarkable and indicates the progressive mindedness of that industry.

Basic principles.

Flotation processes are largely built upon the fact that water wets some minerals rather easily but does not wet others so quickly. The oxides, carbonates, silicates, sulfates and hydroxides are among those that water wets easily. On the other hand, oils and gases adhere to other minerals rather easily while water does not wet them so readily. Among such minerals are the sulfides, arsenides, tellurides, native metals, and carbonaceous materials. The specific gravity (weight of mineral per unit volume) of the mineral, providing it is in a very finely divided condition (48 mesh or finer) plays little if any part when this method of extraction is employed. Very fine sands and slimes have always been the "bug-a-boo" where gravity concentration and amalgamation recovery processes were employed, but these are handled effectively by flotation.

Flotation methods.

Among the various flotation methods that have been developed are those that are commonly referred to as "the surface tension flotation method," "the bulk oil flotation method," and "the froth flotation method." In present-day practice "the froth flotation method" is used in the vast majority of all successful installations.

Froth flotation.

This method is based upon the fact that, when water and oil or any other frothing reagent are violently agitated or churned up together, oil- or reagent-covered bubbles are produced. If finely crushed ore containing sulfides, tellurides, or native metals is treated and if the mixture is violently agitated or churned up, the oil or reagent surface-films of the bubbles attach themselves to these mineral particles and float them to the surface.

Froth producers.

The bubbles making up the froth are produced by either mechanical agitation, pneumatic agitation, pressure reduction, or chemical methods. As the methods using mechanical agitation and pneumatic agitation handle by far the largest tonnages of ore, they are of prime importance.

Mechanical agitation.

Mechanical agitation, as used in flotation processes, consists of the violent churning or beating of the mixture (water, finely crushed mineral, oil, and other reagents) by mechanical beaters or paddles so that air is introduced into the mixture. This action produces bubbles which rise to the surface, forming a froth. Pneumatic agitation is accomplished by blowing air, under pressure, into the above mentioned mixture. This method likewise produces a violent agitation of the mixture and gives results similar to those obtained with mechanical agitators.

Flotation reagents.

The reagents and oils that are used in the flotation process are usually designated as collectors, frothers, depressors and regulators. These consist largely of pine oils, fuel oils, turpentines, wood creosotes, gas oils, organic frothing agents, acids, alkalies, and various salts. The quantity of any single flotation reagent used varies from as little as one two-hundredth of a pound of the reagent to a ton of the ore to as much as five pounds to a ton of ore. There are hundreds of these reagents and new and better ones are being discovered almost daily. They have made it possible to separate not only the sulfides, native metals, etc., from the gangue minerals of an ore, but, also to separate them from each other—for example, lead sulfide from zinc sulfide, molybdenum sulfide from copper sulfide.

Precious metals recovered by flotation.

Although flotation processes revolutionized the extraction methods used on the base metal ores and although, even fifteen years ago, their use in those fields had been brought to a high degree of efficiency, it is only since the discovery and development of the xanthates, aerofloats, minerecs, and other reagents and only within the last five or six years that its use in the treatment of precious metals and their ores has been given serious consideration.

Flour gold and complex ores treated.

In milling, gold ores containing elements that interfere with the economic use of the cyanide process and ores that have their gold values in extremely intimate association with sulfides and tellurides have proven amenable to treatment by this process. It can be used in conjunction with all of the older methods of extraction and it has been proven especially useful in recovering the fine or flour gold that is often lost when only gravity concentration and amalgamation processes are used.

Prospects for small flotation units for gold placers.

As already mentioned, the flotation process has, for several years, been successfully used in the recovery of fine, flour gold from gold ores. Much experimental work has been done on placer gravels and the fine, flour gold has been successfully recovered from those gravels by this process. The only trouble now with its application to small placers is that the process, in its present stage of development, requires so much capital for equipment and such well trained operators that it is of little or no use to the individual placer miner, located here and there, on small workings. Nevertheless, it is a reasonable expectation that, if the same brains, initiative, and energy are concentrated on the development of small flotation units to meet the needs of such men, such equipment will be made available. It will then be possible to increase the gold production of this country markedly since many deposits where flour gold predominates may then be treated profitably and many other deposits will yield higher profits than when only gravity concentration and amalgamation are used, as at present.

APPENDIX TO PART II

SELLING GOLD

The Federal regulations relative to the sale of gold have been changed twice since May 1st, 1933, and it has been thought best not to refer to them herein since it is quite likely that anything written about them would soon be out of date and misleading.

Anyone having gold for sale should apply to one of the United States Mints or Assay Offices for the current regulations relative to the sale of gold.

The United States Mints are located at Philadelphia, Pennsylvania; San Francisco, California; and Denver, Colorado.

The United States Assay Offices are located at 32 Old Slip, New York City, and at Seattle, Washington.

BIBLIOGRAPHY

Boericke, W. F., Prospecting and Operating Small Gold Placers: John Wiley & Sons, Inc., New York, 1933.

Bowie, A. J., Jr, Practical Treatise on Hydraulic Mining, 11th ed.: D. Van Nostrand Co., New York, 1910.

Gardner, E. D., and Johnson, C. H., Placer Mining in the Western United States: U. S. Bureau of Mines Bulletin in preparation.

Idriess, I. L., Prospecting for Gold, 3rd ed.: Angus & Robertson, Ltd., Sydney, Australia, 1932.

Longridge, C. C., Hydraulic Mining: Mining Journal, London, 1910.

Peele, R., Mining Engineer's Handbook, 2nd ed.: John Wiley & Sons, Inc., New York, 1927, pp. 882-988.

Raeburn, C. and Milner, H. B., Alluvial Prospecting: D. Van Nostrand Co., New York, 1927, pp. 126, 130, 141.

Taggart, A. F., Handbook of Ore Dressing: John Wiley & Sons, Inc., New York, 1927, pp. 639-649.

Voll, M., A.B.C. of Practical Placer Mining: Great Western Publishing Co., Denver.

Von Bernewitz, M W., Handbook for Prospectors: McGraw-Hill Book Co., New York.

Wilson, E. B., Hydraulic and Placer Mining, 3rd ed.: John Wiley & Sons, Inc., New York, 1918.

Young, G. J., Elements of Mining, 3rd ed.: McGraw-Hill Book Co., New York, 1932, pp. 424-467.

U. S Geological Survey Publications:

Brooks, A. H, Genesis and Classification of Placers: Bulletin 328, 1908, p. 111.

Brooks, A. H., The Mineral Deposits of Alaska: Mineral Resources of Alaska, 1913, 1914, p. 413.

Day and Richards, Investigations of Black Sands from Placers: Bulletin 285, 1906.

Jones, E. L., Jr., Gold Deposits near Quartzsite, Arizona: Bulletin 620, p. 45.

Hutchins, H. P., Prospecting and Mining Gold Placers in Alaska: Bulletin 345, p. 54.

Purington, C. W., Methods and Costs of Gravel and Placer Mining in Alaska: Bulletin 263, 1905.

U. S. Bureau of Mines Publications:

Jackson, C. F, and Knaebel, J. B., Small-Scale Placer Mining Methods: Information Circular 6611, 1932.

Janin, C., Gold Dredging in the United States: Bulletin 127, 1918.

Jennings, Hennan, The History and Development of Gold Dredging in Montana, with a chapter on placer mining methods and operating costs by Charles Janin: Bulletin 121, 1916.

Wimmler, N. L., Placer Mining and Costs in Alaska: Bulletin 259, 1927.

State Publications

California.	Gold Placers of California—Mining Methods, by C. S. Haley: California State Mining Bureau, Bulletin 92, 1923. Mining in California, by C. M Laizure and others, 1932: Vol. 28, No. 2, Reports of the State Mineralogist, Division of Mines, San Francisco.
Idaho.	Elementary Methods of Placer Mining, by W. W. Staley: Idaho Bureau of Mines and Geology, Moscow, Idaho, 1932.
Montana.	Placer-Mining Possibilities in Montana, by O. A. Dingman: Montana School of Mines, Butte, Memoir No. 5, 1932.
Nevada.	Placer-Mining in Nevada, by W. O. Vanderburg and A. M. Smith: State Bureau of Mines, Reno, 1932.
South Dakota.	Prospecting for Placer Gold in South Dakota, by D. L. M. Anderson: State Geological Survey, University of South Dakota, Vermillion, S. Dak., 1933. Geologic History of Black Hills Gold Placers, by J. P. Connolly: State Geological Survey, University of South Dakota, Vermillion, S. Dak., 1933.
Washington.	Small Scale Methods of Placer Mining and Placer Mining Districts of Washington and Oregon, by G. E. Ingersoll: School of Mines, State College of Washington, Vol. 15, No. 6, 1933.
British Columbia.	Placer Mining in British Columbia, by J. D. Galloway: Department of Mines, Victoria, 1931.

Tables and Conversion Data

TROY WEIGHTS AND EQUIVALENTS

The troy grain, ounce and pound weigh the same as the respective weights of the apothecaries' system and are used for weighing gold and silver. The troy ounce is NOT the same as the avoirdupois ounce; nor is the troy pound the same as the avoirdupois pound. To convert avoirdupois ounces into troy ounces, multiply by 0.911. To convert troy ounces into avoirdupois ounces, multiply by 1.097. In other words, a troy ounce is about 10 per cent heavier than an avoirdupois ounce. To convert troy pounds into avoirdupois pounds, multiply by 0.8229, and, to convert avoirdupois pounds into troy pounds, multiply by 1.125. The avoirdupois pound is, therefore, about 12 per cent heavier than a troy pound.

The following is the troy system of weights and measures, together with some useful equivalents.

Troy Weights and Measures.

24 grains = 1 pennyweight.
20 pennyweights = 1 ounce = 480 grains.
12 ounces = 1 pound = 240 pennyweights = 5,760 grains.

Equivalents.

1 ounce, troy = 1.097 ounce, avoirdupois = 31.103 grams.
1 ounce, avoirdupois = 0.911 ounce, troy = 28.35 grams.
1 pound, troy = 0.8229 pound, avoirdupois = 13.166 ounces, avoirdupois = 373.2 grams.
1 pound, avoirdupois = 1.215 pound, troy = 14.58 ounces, troy = 453.6 grams.

Liquid volume and capacity equivalents.

1 U. S. quart = 2 pints = 0.25 U. S. gallon = 57.75 cubic inches = 0.334 cubic foot = 0.00397 barrel = 0.946 liter.
1 U. S. gallon = 8 pints = 4 quarts = 231 cubic inches = 0.1337 cubic foot = 0.0317 barrel = 3.785 liters.
1 barrel = 31.5 U. S. gallons = 126 U. S. quarts = 252 pints = 7,276 cubic inches = 4.21 cubic feet = 0.119 cubic meter.
1 cubic foot = 1,728 cubic inches = 0.237 barrel = 7.48 U. S. gallons = 29.92 U. S. quarts = 28.3 liters.

WEIGHTS OF MATERIALS

Name	Pounds per cubic foot	Cubic feet per short ton of 2,000 lbs.
Barium sulfate (barite)..	280	7.1
Basalt (trap-rock	181	11.0
Brass (copper and zinc) cast	527	3.7
Copper, cast	550	3.6
Copper iron sulfide (chalcopyrite)	262	7.6
Diabase or diorite	187	11.4
Gold, native	1185 approximately	1.7 approx.
Granite	170	11.7
Gravel, wet	125 approximately	16.0 approx.
Iron, cast	450	4.4
Iron sulfide (pyrites)........	318	6.3
Lead carbonate (cerussite)	409	4.9
Lead, cast	711	2.8
Lead sulfate (anglesite)..	388	5.1
Lead sulfide (galena)........	467	4.3
Magnetite (black sand)....	318	6.2
Mercury, (quicksilver), native	898	2.2
Mica	183	10.9
Porphyry	162	12.4
Quartz	162	12.4
Sand ....90 to 130		15.4 to 22.2
Silver, native, average....	655	3.0
Steel, cast	492	4.0
Tin oxide (cassiterite)......	424	4.7
"Tungsten" (scheelite)....	455	4.4
Water	62.5	32.0

PART III

SUGGESTED LIST OF EQUIPMENT FOR PROSPECTING IN THE SOUTHWEST[1]

By Charles H. Johnson

Formerly Assistant Mining Engineer, Southwest Experiment Station, U. S. Bureau of Mines

The following list of equipment is intended as a check list to aid in securing the essentials. It is not complete, as clothing or other personal effects will conform to each person's ideas. Some items may be omitted for economy or lack of space, or because they are not needed under certain conditions.

Prospecting tools.

Hammer (one single jack, about 4-pound weight), shovel (round point, long handle), miner's pick, prospector's pick, moils (two or three), gold pan, horn-spoon, small mortar and pestle, magnifying glass, blow-pipe outfit, determinative tables, sample sacks (some use double paper sacks), compass, maps (topographic and geologic).

If the prospector expects to make a permanent camp and do some hard rock mining:

Powder, caps, fuse, hand steel, spoon (to clean drill holes), tamping stick, and blacksmithing equipment and tools.

General camping equipment.

Tent, tent pins (steel), lengths of rope, canteens (one small and one large), axe, saw, hammer, assorted nails, folding cot (to keep bed off ground), blankets, canvas blanket cover and wrap, water containers (5 gallon gas cans or kegs), pail, soap, lantern or acetylene lamp and carbide, flashlight, matches, jack-knife, and canvas bags in which to stow clothing and other things.

Cooking equipment.

Large stew pan (one or two), small stew pan (one or two), grill, frying pans (one large and one small), large iron spoon, carving knife, can opener, Dutch oven, knives, forks, spoons, tin, aluminum, or enameled ware (cups, plates, coffee pot, etc.), towels, etc.

Medical and first-aid supplies.

Take along your own medicines—those that you are accustomed to use. Be certain that they include a laxative, an emetic, iodine or mercurochrome, and a first-aid kit. Castor oil, salts, or backing soda may be used as laxatives or purgatives. Large quantities of salt water or mustard water serve as emetics (to cause vomiting). Many other equally effective remedies are available. A snake-bite kit is advisable. (See Part VI).

---

[1] Published by permission of the Director, U. S. Bureau of Mines.

PART IV

FOOD SUGGESTIONS FOR A PROSPECTOR

By Margaret Cammack Smith

Professor of Nutrition, University of Arizona

Do you know that how you feel depends largely upon what you eat every day? Your ability to stand the heat of the sun, your ability to work without undue fatigue, your ambition and enthusiasm depend upon what you eat perhaps more than upon any other one thing.

The good health of many a pioneer and prospector was undermined by wrong food. A diet of beans and bacon, soda biscuits, and such foods does not make continued good health possible. You must have a well rounded, balanced food supply.

Foods differ in composition and, therefore, in their effect and value to the body. They are classified as follows:

1. **Energy foods.** Those foods that supply the fuel for the work of your muscles. An insufficient amount of energy-giving foods results in inefficiency and in loss of body weight. The body burns its own fat and muscle tissue unless enough food is eaten. Cereals, starchy foods, fats, sweetstuffs, and dried legumes (beans and peas) are the chief providers of energy.

2. **Muscle-building foods.** These are necessary for building muscles and keeping already formed muscles in repair. Meats, fish, eggs, cheese and nuts are muscle builders.

3. **Protective foods.** Milk, eggs, vegetables, and fruits are in this class. Special attention should be paid to these foods since the minerals and vitamins which they contain protect you against disease. They aid in the prevention of nervous disorders (neuritis, beriberi, etc.), digestive upsets, scurvy, tooth decay, anemia, rheumatism, tuberculosis, colds, and other infections of sinuses, nose, throat, and lungs.

4. **Bulky food—Roughage.** Certain foods are necessary to keep the system lubricated, to promote the proper action of the digestive tract and bowels, to aid in the elimination of waste products. Constipation and its accompanying ills, such as sluggishness, headaches, etc., can be prevented by the liberal use of the bulky foods, especially fruits, vegetables and whole grains.

The weekly food allowance given below has been planned to satisfy the food needs for the maintenance of health of a workman. As far as possible, only foods which can be kept for long periods of time in good condition without ice have been included.

Group A (Milk)

Food	*Amount*	*Remarks*
Milk	*At least* three 1 lb. cans of evaporated or 1 lb. dried milk (Klim).	This amount is the equivalent of 3 qts. of fresh milk. Milk is the most essential food for it is the *greatest* health protector. More milk might better be used and it can be used in any way desired.

Group B (Vegetables)

Carrots Cabbage Onions Beets Parsnips Turnips Squash Potatoes String beans Peas Spinach Corn	At least 6 lbs. of vegetables. Not more than 2 lbs. of potatoes should be included.	If fresh vegetables, such as lettuce, spinach, celery, asparagus, string beans, etc, are available, they should be used. Canned vegetables are more expensive but otherwise, weight for weight, can be safely substituted. Greens, such as spinach, lettuce, cabbage, beet greens, or alfalfa, are excellent.

Group C (Fruits)

Prunes Apples Peaches Apricots Dates Figs Raisins	1 lb. of dried fruits.	
Canned tomatoes or Citrus fruits or Other fresh fruits	1 lb. can 3 6	Canned tomatoes or citrus fruit or some *fresh raw* fruit or vegetable is necessary each day to prevent scurvy, rheumatism, etc.

Group D (Cereals and legumes)

Dried beans or peas Rolled oats Wheatena Shredded wheat Cracked wheat Brown rice Whole wheat flour Whole wheat bread Cornmeal	2½ to 3 lbs. 6 to 8 lbs. of whole grain products.	These are the energy foods. They satisfy hunger and provide fuel for work.

Group E (Meat or Meat Substitutes)

Dried or canned meat or fish Bacon Ham Cheese	2½ lbs, from this group.	These foods are the muscle builders. Fresh meat or eggs may be used if available.

You will also want some sugar, coffee, tea, salt, and baking powder, and perhaps some fat, such as ½ lb. of butter, salt pork, canned lard or Crisco to make the food ration more palatable.

Remember to Eat for Health.

Be especially careful to eat enough of the health protecting foods, namely, *milk, fruits, and vegetables.* For the most part these foods may be canned or dried, providing the fresh food is not available. A good scheme to follow is to see that every day you eat:

1. One pint of milk or its dried or canned equivalent used in any way desired.

2. Two fruits. One of these should be a fresh raw fruit or canned tomatoes or citrus fruit.

3. Two vegetables. If possible, one of these should be a green vegetable of some kind.

4. Meat, cheese, fish, or eggs, once a day.

5. A whole grain cereal or breadstuff.

Then satisfy your appetite with more of the same foods or with other foods, as you desire.

PART V

TREATMENT OF SUNSTROKE OR HEAT PROSTRATION

By Fred P. Perkins, M. D.

Medical Advisor, Director of Health, University of Arizona

Treatment in the Field

When a case of heat prostration or sunstroke develops in the field, first remove all tight clothing and put the patient in the shade. Dash cold water on head and body at frequent intervals, not continuously. The throwing of water has a two-fold purpose: cooling the surface and stimulating the cardio-vascular system by the force of the water being dashed upon the surface, causing a contraction and subsequent dilitation of the capillary circulation, forcing the cooled blood away and the hot blood to the surface. Give frequent sips of cool water internally and move patient to hospital as soon as possible.

Treatment in a Hospital

The best results have been obtained by following the method used at St. Vincent's Hospital in New York where a series of 197 cases were treated with a mortality of only 6 per cent. The patient is immediately wrapped in a sheet and placed on a cot which has been covered with a rubber blanket. Dippers full of cold water are dashed upon him from a distance of several feet. Every two or three minutes a stream of ice cold water is poured on his head from a height of six or eight feet. When his temperature falls to 103° F., he is wrapped in a blanket and surrounded with hot water bottles. The gradual reduction of temperature proves the safest method. When the patient is wrapped in a sheet upon which the water is dashed, a too sudden withdrawal of the heat from the surface is prevented. Unfavorable symptoms have developed when ice tubs or packs have been used.

This treatment must be given carefully and the patient watched continuously for sudden changes in the circulation; and it may be necessary to give some quick heart stimulant, such as tincture of camphor or aromatic spirits of ammonia, to tide over a weak heart for a time. Always remember the patient's condition, a superheated condition, and don't lose your head and overtreat. For excessive thirst, have patient put something in his mouth which will stimulate the salivary glands—a pebble or small piece of wood or chewing fum will often be all that it necessary.

PART VI

INFORMATION ON POISONOUS ANIMALS

By C. T. Vorhies

Professor of Entomology, University of Arizona

Scorpion and Centipede

Stings of the common scorpion are not dangerous to adult persons. In most instances and for most individuals, the scorpion sting is less painful than the sting of a common honey bee. It is sometimes dangerous for small children, however.

The bite of the large centipede, while very painful, is not really dangerous. The centipede has a single pair of poison jaws under the head. The claws of the remaining legs are not poisonous and are not capable of leaving a fiery trail, notwithstanding the many stories to the effect.

Weak ammonia may be somewhat helpful as a palliative for scorpion stings and centipede bites, though I do not feel at all certain that it accomplishes anything. Strong ammonia, on the other hand, will produce a burn more severe and requiring a longer period for recovery than the original sting or bite. I believe that suction is of value in the case of all poisonous stings or bites.

Tarantula

The tarantula, while generally very much feared, is not so apt to bite as is usually believed and is not very painful or dangerous when it does so. Secondary germ infection of all such wounds is more or less likely to occur and should be guarded against by antiseptic treatment.

Vinegarone, etc.

The vinegarone, child of the earth, praying mantis, and similar forms sometimes regarded as dangerous are in fact nonpoisonous and entirely harmless.

Black Widow Spider

The black widow or shoe-button spider is a really poisonous species. It is usually jet black, shiny, the size of a pea or old-fashioned shoe button and has an hour-glass shaped red mark beneath the body. Illness, sometimes serious, frequently follows the bite of this spider. Fortunately, however, persons are not bitten by the black widow as often as might be expected since the spider is really quite common, in southern Arizona at least. A physician's care is needed in case of such a bite. I make no recommendations as to treatment.

GILA MONSTER

The Gila Monster is the only venomous lizard in the world. All others are harmless so far as poison is concerned. The Gila Monster is not deadly. There is no absolutely authentic case of human death certainly due to Gila Monster bite on record, all stories to the contrary notwithstanding. Let these animals alone and you will never be bitten, as may be the case with rattlesnakes. I know of no case where a bite has been inflicted except where the victim was fooling with or teasing the Gila Monster. In case a bite is incurred, release the bitten part as soon as possible. Wash the wound and apply suction. The wound should be treated with antiseptic. See a physician if symptoms develop.

SNAKES

Rattlesnakes are the only really dangerous poisonous snakes which we have to fear in Arizona. The coral snake, while its venom is doubtless more deadly, drop for drop, than that of the rattlesnake, is too small to be regarded as very dangerous and there is no fatality on record for this species as far as I know. In fact I do not even have a record of any case of a bite by this species in Arizona.

Treatment of rattlesnake bite. A full account of treatment of rattlesnake bite goes rather beyond the bounds of this statement, but perhaps I can outline the most important things within such limits First of all, discard the idea of using potassium permanganate. Experiments have proved it is worthless in any strength that one would dare use.

Let us divide the treatment into two phases: first, emergency treatment or first aid, and second, medical care. Every case of rattlesnake bite should be brought to the care of a physician as soon as possible, so you are concerned primarily with the first phase.

1. *Keep cool*, rattlesnake bites are *painful*, but only a small percentage is fatal.

2. Apply tourniquet at once between wound and heart, tight enought to hinder venous circulation, not necessarily tight enough to shut off arterial flow. A stout band or strip of rubber is good and can be most quickly applied.

3. Open fang punctures by cross cuts ⅛ inch deep, with sharp, sterile knife. (Safety razor blade is easily carried in sterile package). Suck the wound, by mouth if necessary. Mouth must be free of wounds or abrasions. Best to have a suction bulb, or apply suction mechanically as soon as possible, since long continued suction has been proved efficacious.

4 Loosen tourniquet every 20-30 minutes for two or three minutes.

5. If *Antivenin* be at hand, administer at once according to directions.

6. Keep patient *quiet*. Give stimulant if there is weak heart action or fainting. *Alcohol is not a stimulant.* Black coffee, aromatic spirits of ammonia, and strychnine, *are* stimulants. (Plenty of alcohol will neatly finish what the venom has started.)

7. As soon as possible get a physician who should continue the suction treatment, give *Antivenin,* if not previously given, and care for the wound to prevent infection.

Suction removes venom in bloody serum for hours after bite is inflicted. *Antivenin* counteracts the venom which has gotten into the blood stream, and will benefit many hours after bite. Be careful not to *slash* indiscriminately or too deeply in opening for suction and drainage, especially on hand, foot, wrist, or ankle, as serious damage to tendons may result.

Incision and suction have been successfully used in Texas in cases of rattlesnake bite. The removal of poison in blood or bloody serum from the wound may be increased by strong suction, and for many hours after the bite is inflicted. Suction should be kept up for twenty minutes out of each hour over 'a period of fifteen hours or until swelling ceases. Mechanical suction is, of course, necessary for this purpose. We can recommend the "B-D" outfit costing $1.50. Many cases of rattlesnake bite could be saved by proper use of this treatment alone, though we believe in *Antivenin* as an additional precaution and relief. The physician (who should be seen as soon as possible) should have, or be able to secure, *Antivenin.*

Very few bites are quickly fatal. Most of the fatal bites do not result in death before 18 to 48 hours so there is ample time for treatment.

DON'TS

Don't run or get overheated. Circulation, increased by exercise or by alcohol, serves to distribute the poison much more rapidly through the body. Don't injure the tissues by injecting potassium permanganate, which is now known to be of no value as an antidote. Do not depend upon snake bite "cures" or home remedies commonly used. They are of no value. Do not cauterize the site of the bite with strong acids or the like.

Don't forget *strong black coffee,* and don't take whiskey or other alcoholic drink.

PART VII

LAWS, REGULATIONS, AND COURT DECISIONS IN
RELATION TO THE LOCATION AND RETENTION
OF GOLD PLACER CLAIMS IN ARIZONA

By G. M. Butler

Director, Arizona Bureau of Mines

INTRODUCTION

Although occurrences of many substances besides gold or
other precious metals may be located as placer claims, which
substances include alum, asphalt, borax, diamonds, guano, gyp-
sum, kaolin or china clay, marble, mica, onyx, various salts of
soda, roofing slate, umber, building stone, etc., every statement
made in this bulletin about locating and holding placer claims
should be understood to apply specifically to claims chiefly valuable for the gold, silver, or other precious metal which they
contain.

Since the Federal and State statutes bearing on the matters
discussed are not always entirely clear, it has frequently been
necessary to appeal to the courts for decisions on various questions relating to placer claims. Sometimes the decisions of different courts seem to be at variance with each other, and one
cannot speak positively about the true intent of the law in such
cases.

The statements that follow are either transcriptions of the
statutes themselves or are based on court decisions. It is believed by the writer that they are trustworthy, but he recog-
nizes that some of them are merely matters of opinion based on
court decisions which might be reversed by higher courts.

WHO MAY LOCATE PLACER CLAIMS

A citizen of the United States (either male or female) or one
who has declared his intention to become a citizen before the
proper court may locate placer claims.

An Indian or a minor child may make valid locations as may
an unincorporated group or association of people who individ-
ually are entitled to locate placers. Locations may also be made
by a corporation chartered under the laws of any state or ter-
ritory of the United States, but no officer or employee of the
General Land Office, including clerks, special agents, and mineral
surveyors, can locate such claims.

Size and Shape of Placer Claims

Each placer claim, if located by *one* person or a corporation (a company incorporated under the laws of any state, which is legally an individual), may not exceed twenty acres in area and should conform in shape to the system adopted in surveying the public lands of the United States. If previously located claims or patented ground cover part of the area desired, however, such property may be excluded from a placer claim even if to do so gives the claim a non-rectangular shape and necessitates running one or more of the boundaries in a direction other than north and south, or east and west.

A 20-acre placer claim should cover half of a quarter of a quarter section if located on surveyed land and measure 660 feet by 1,320 feet. The smallest tract recognized or considered is 10 acres and measures 660 feet by 660 feet. Five-acre tracts will not be recognized. If a claim is located on unsurveyed land, it should have the same dimensions and shape that it would have if on surveyed land and the boundaries must run north and south, and east and west (true, not magnetic, bearings).

While it has been held that, where the topography is such as to make it practically impossible to lay out rectangular claims, as in gulches with precipitous walls, placer claims may be located so as to conform to their environment, the land office regulations require that entries be as compact as possible in form and this office will not approve entries that cut the public domain into long, narrow strips or decidedly irregular tracts. The locators of placer claims will, therefore, probably find it advantageous to take up only square (10 acres) or rectangular (20-acres) claims of the dimensions already given if it is possible to do so. An individual or a corporation may locate any number of placer claims.

The Location of Claims by Groups of People

The law permits *two* individuals to locate a 40-acre placer claim, *three* persons a 60-acre claim, etc., but the largest placer claim that can be located by a group or association of persons is 160 acres, and *eight* people must constitute the locating group for a claim of that size since not over 20 acres is allowed for each individual constituting such a group. The provision that allows 20 acres for each member of a group not exceeding eight persons has been the basis of rank fraud. One person may locate the claim and use the names of other people as dummies who really have no interest in the ground, who afterwards convey their interest to the real locator thus enabling one person really to locate and eventually to acquire a placer claim that may contain as much as 160 acres. Such locations, if proved to be fraudulent, will be set aside by the courts.

LOCATION PROCEDURE

The first requirement is the discovery of mineral within the limits of the claim to be located. The grade of the metal-bearing sand, gravel, loose rock, or dirt does not have to be such as to make it possible to work it profitably where the discovery is made, but it does have to contain such values as would encourage a prospector to do more work at or near the point of discovery in the hope of finding material of higher grade.

When a claim is located by a group or association of two to eight individuals, the discovery of mineral at a single point is all that is required.

After the discovery of mineral, the following things must be done:

1. Post a location notice.
2. Mark the boundaries of the claim, at each angle, with a post or monument of stones.
3. Within 60 days after the date of the location, record a copy of the location notice in the office of the county recorder.

THE LOCATION NOTICE

Arizona statutes require that the location notice must contain:
a. The name of the claim.
b. The name (or names) of the locator (or locators).
c. The date when the claim was located.
d. The number of acres claimed.
e. A description of the claim with reference to some natural object or permanent monument that will identify the claim.

Blank location notices or certificates can be obtained from stationery stores for five cents each. The following blank notice, however, may be used.

...., a citizen (or citizens) of the United States (or who has—or have—declared his—or their —intention—or intentions—to become a citizen—or citizens—of the United States), the undersigned, claims (or claim) by right of discovery and location this, the..placer mining claim which contains..acres, situated in the..Mining District, County of ...., State of Arizona, which is bounded and described as follows, to wit:

Beginning at (if on surveyed land, make the starting point a corner of one of the subdivisions of such survey).., a (post or stone monument), where this notice is posted, hence (give direction)............................feet to a (post or stone monument), hence (give direction)............................feet to a (post or stone monument, hence (give direction)............................feet to (a post or stone monument), hence (give direction)............................feet to the place of beginning from which (mention some natural object or permanent monument) bears (give approximate direction and distance).

Dated and posted on the ground this..day of .... 19.........

Signed..

....

The law does not require that the names signed to a location notice be signatures. In other words, anyone can sign for someone else. The location notice should be prominently posted at some point on the claim, as at the point of discovery or one of the corner posts. The printed blank placer location notices, as well as the blank just given, assume that it is posted at one of the corners of the claim. It should be posted in such a way as to be protected from the weather, and a good plan is to tack it on the inside of a stout, tight box which has been nailed to a post.

MARKING BOUNDARIES

Arizona statutes require that a post or monument of stones be-erected at each angle of a placer claim.

If wooden posts are used, they must be at least four inches in diameter and four feet six inches in length, and each post must be set at least one foot into the ground and surrounded by a mound of earth or stone of unspecified height. Where it is impossible to set post into the ground, they may be supported by piles of stones.

When a mound of stones alone is used as a monument, it must be at least three feet high and four feet in diameter at the base.

If it is impossible to erect and maintain a post or monument of stones, a witness post or monument may be used, to be placed as near the true corner as the nature of the ground will permit.

No Discovery Excavation Necessary on Placer Claims

Although gold or some other valuable mineral not in a vein or lode must be discovered before a placer claim may be legally located and although considerable excavation work must sometimes be done to make such a discovery, the Arizona or Federal Laws do not require that any work be actually done on a placer claim in order to complete the location procedure. In other words, no work not needed to make the discovery of mineral need be done on a placer claim until after noon of July 1 following its location.

If a placer claim exceeds 20 acres in size, the work mentioned may be all done at one point.

Annual Labor or Assessment Work

In order to retain an unassailable title to placer ground which has been legally located, $100 worth of improvements must be made or $100 worth of work must be done on each *claim*, regardless of its size, every year, between noon of July 1 following the location of the claim and the next following July 1 at noon and during each succeeding year until a patent is obtained. This work may all be done at one point no matter what the size of the claim. If the annual labor has been commenced, but has not been completed by noon of July 1, and is diligently carried on thereafter until completed, the owner of the claim is deemed to have complied with the law, but an additional $100 worth of work must be started or completed before noon of the next succeeding first of July in order to retain title to the property.

PERMISSIBLE ASSESSMENT WORK ON PLACERS

The Federal or State statutes do not specify exactly what work will satisfy the requirements of the annual labor law. Court decisions furnish the only guide in these matters, and they, naturally, do not cover all of the questions that might be propounded. Such questions can be answered positively only by carrying the matter through the courts, and one man's opinion is as good as another's.

From court decisions, which other courts might reverse, we learn that the cost of doing certain things on or in connection with placer claims may be considered as probably fulfilling the annual labor provisions of the law, as follows:

1. Digging of a prospect hole or holes or of a cut or cuts.
2. Digging of a drain ditch or ditches.
3. Removing brush so as to get at the underlying gravel.
4. Constructing ditches, flumes, and pipe lines for conducting water to the claim for use in mining thereon.
5. Erecting of other works for mining.
6. Installing machinery.
7. Constructing any building, roadway (built *exclusively* in order to benefit the claim and necessary in order to develop it), and other improvements used in connection with and essential to the development of the claim.
8. Making drill tests on placer ground, if done in connection with actual dredging operations on adjoining claims.
9. Purchasing a dredge and placing it on the claim.
10. Employing a watchman to take care of and protect mining property while idle if such services are necessary to preserve buildngs or other structures erected to work the mine, which would be needed if actual work were resumed, provided it is intended to use such structures again within a reasonable time.
11. Constructing dams or reservoirs on or off the property for the sole purpose of storing water to be conducted to and used in mining operations on the property.
12. Constructing a flume, which may be partially off the property, to carry away tailings and other waste material.

WORK THAT MAY NOT BE USED AS ANNUAL LABOR

Court decisions (which likewise might be reversed by other courts) have been made to the effect that the cost of doing certain things on or in connection with placer claims do *not* fulfill the annual labor provisions, as follows:

1. Constructing a reservoir on a claim to store water to be conducted and used elsewhere.

2. Expending money in travel in an endeavor to arrange to conduct water to the claim.

3. Placing on the ground tools, implements, lumber, and other material which are not used to any extent and are subsequently removed.

4. Erecting log cabins to be used by laborers (other decisions have approved the construction of necessary buildings, such as boarding houses, bunk houses, stable, and blacksmith shops as fulfilling the requirements of the annual labor law).

5. Erecting a stamp mill or a lime kiln.

6. Employing a watchman merely to warn off prospectors and to prevent claim jumping or to prevent the stealing of small tools like picks, shovels, etc.

7. Surveying for a ditch, flume, or pipe line if the ditch, flume, or pipe line has not been dug or constructed.

FILING AFFIDAVIT OF PERFORMANCE OF ANNUAL LABOR

Neither the Arizona nor the Federal statutes require that affidavits that the annual labor has been performed *must* be filed with anyone, but the Arizona statutes provide that such affidavits *may* be filed with the Recorder of the County wherein the claim is located, within three months after the expiration of the time fixed for the performance of annual labor or the making of improvements upon a mining claim. In other words, under existing laws, such affidavits must be filed prior to noon of the first of October that follows the first of July after the work is completed, if filed at all.

The Arizona statutes also provide that the recording of an affidavit of annual labor constitutes *prima facie* evidence that the labor has been completed. It is, therefore, desirable to file such affidavits in order that recorded evidence that the title to the claim is clear may be available.

Blank affidavits of performance of annual labor may be secured at stationery stores.

LODES WITHIN PLACER CLAIMS

A placer location does not cover any lodes that may exist within it. Either the holder of the placer claim or any stranger may locate and hold a lode included within an unpatented placer claim by following the procedure laid down in the Federal and State statutes for locating lode claims, just as though no placer claim had been located around the lode. Furthermore, strangers are permitted to enter peaceably upon an unpatented placer claim, without the owner's consent, to locate a lode claim. After a placer claim has been patented, its owner also owns any lodes included within it if their existence was unknown when the patent was issued.

PLACERS WITHIN LODE CLAIMS

The owner of a lode claim is also the owner of any unlocated placer deposits that may exist thereon. It is, however, essential that the lode claim should have been legally located on a deposit of mineral in place, in vein or lode form, and that all requirements of the statutes that govern the acquirement and holding of lode claims shall have been met.

PLACERS ON INDIAN RESERVATIONS

Mining claims, lode or placer, cannot be located and held on Papago Indian Reservations.

According to the act of June 30, 1919, citizens of the United States, associations of citizens, or domestic corporations may prospect for deposits of gold, silver, and other valuable metalliferous minerals on unallotted lands on Indian reservations, and if a deposit is found, may locate and lease it from the Government, if the Secretary of the Interior has declared that the area where the deposit is found is open to exploration.

Placer mining claims are located and held on all Arizona Indian reservations, except Papago reservations, exactly as on the public domain in that state, with the following exceptions:

1. Persons who have merely declared their intention to become citizens of the United States cannot make valid locations.

2. A copy of the location certificate must be filed with the Superintendent of the Indian Reservation, and the County Recorder, within sixty days.

3. After the claim has been located, the locator has one year within which to apply for lease through the Superintendent of the Reservation. Such a lease is for twenty years, and the holder has preferential rights to obtain renewals for successive 10-year periods.

4. After the lease has been obtained, the holder must pay monthly a royalty of not less than 5% of the net value of the output from the claim. At the time this bulletin goes to press, the royalty on the net value of the output from gold deposits is fixed at 10%.

5. After the lease has been obtained, the claim holder must pay *in advance* an annual rental of at least 25c per acre for the first calendar year after the lease is obtained, at least 50c per acre for the second, third, and fourth calendar years, respectively, and at least $1.00 per acre for each year thereafter during the term of the lease. The rental is credited against the royalties as they accrue.

6. A locator on unsurveyed land must at his own expense have the claim surveyed by a U. S. surveyor before a lease will be granted.

7. Not over 40 acres may be leased for not less than $1.00 per acre for a camp site, mill site, smelting and refining works, etc., by each holder of a mining claim lease.

8. Lands containing springs, water holes, or other bodies of water needed or used by the Indians for watering livestock, irrigation, or water power purposes are not open to exploration and lease.

9. Claims on Indian reservations cannot be purchased.

It should be noted that the usual $100 worth of annual labor must be performed on each claim in addition to the payment of royalty and rental.

Anyone who plans to prospect and mine on an Indian reservation should obtain a copy of the law and the rules pertaining thereto from the Commissioner of the General Land Office, Washington, D. C. A complete set of the necessary application and other blanks can be obtained from him for $1.00.

Since the Secretary of the Interior may at any time declare additional land open to exploration or withdraw land previously open to exploration, it would be wise for a prospector on an Indian reservation to secure from the Superintendent of the Reservation information as to what lands are and are not open to exploration, before he does any work on that reservation.

On May 15, 1932, all of the Wallapai, Western Navajo, San Juan, Navajo, Kaibab, San Carlos, and Fort Apache reservations, and parts of the Salt River, Colorado River, and Gila River reservations had been declared to be open to prospectors and prospective leasers, in Arizona.

PLACERS ON STATE LANDS

When Congress granted statehood to Arizona, it gave the public land in Sections 2, 16, 32, and 36 in each township to the state for school purposes. The state also owns much additional land, and it is necessary to correspond with the State Land Department at Phoenix to learn whether any area is or is not on state land.

Mining claims may be located on Arizona state lands, even such lands as have been leased for agricultural purposes (since the state reserves to itself the minerals on such leased lands), in exactly the same way as though they were on the public domain, with the following exceptions:

1. If a placer deposit is located on state lands not already leased as mineral ground to some one else, the locator should post a location notice, stake out the claim, and apply to the State Land Department for a two-year prospecting permit or lease. A fee of $2.00 must be sent with the application. When

notified that the application has been granted, a rental fee of $5.00 per claim and in issuance fee of $2.00 must be sent to the State Land Department before the permit of lease is delivered.

2. The fact that land has been leased for agricultural or grazing purposes does not prevent a prospector from searching for mineral deposits thereon. He has the right to prospect on such land so long as he does not harm the lease-holder's crops or improvements, and if he finds a mineral deposit, he may locate it and apply for prospecting permit or lease, as stated in the preceding paragraph.

3. It is not necessary to file a copy of the location certificate with the County Recorder.

4. Not more than fifty tons of ore can be removed from a claim held under a prospecting permit.

5. If a commercially valuable deposit is found on land held under a prospecting permit within two years from the date of that permit, and the owner desires to exploit it, he must surrender his permit to the State Land Department and obtain a five-year operating and development lease.

6. The Arizona statutes provide that the annual rental to be paid for an operating and development lease shall not be in excess of five per cent of the net value of the output of minerals. The net value is computed by subtracting all costs, excepting the expense of prospecting and such preliminary work, from the gross value of the ore.

7. Annual labor to the value of five dollars per acre must be performed on placer claims held under lease from the state, and an affidavit that such labor has been performed *must* be filed each year with the State Land Department.

8. The lessee may cut and use timber upon the claim for fuel, buildings required in the operation of any mines on the claim, and necessary drains, tramways, and mine timbers, but for no other purpose.

9. Mineral claims on state land cannot be purchased.

For further information concerning the leasing of mineral deposits on state land and for the necessary forms, applications should be addressed to the Commissioner of the State Land Department, Phoenix, Arizona.

PATENTING PLACER CLAIMS

A placer mining claim, located on the public domain, may be patented or purchased from the United States Government, after at least $500 worth of work has been done upon it, for $2.50 per acre or fraction of an acre plus various fees. If on unsurveyed land, the claimant must also pay for a survey made by a U. S. Mineral Surveyor. Patent procedure is so complicated that it would serve no good purpose to outline it in this bulletin, and anyone who contemplates applying for a patent should consult an attorney-at-law.

INDEX

www.ingramcontent.com/pod-product-compliance
Lightning Source LLC
Chambersburg PA
CBHW032002190326
41520CB00007B/332